THE MELONHEAD GROWERS GUIDE

NATURES PHARMACY

FROM SEED TO HARVEST

CANNABIS CULTIVATION

ORGANIC NO TILL

By Chef Derek Butt

Dedication

I wrote this book in honor of my parents and my younger sister, who I lost to cancer. If I knew then what I know now, they might be still alive today. Unfortunately, their trust for their doctors and their respect for the law is the only reason why I believe they are not alive today.

Copyright 2016.
All rights reserved.
Chef Derek Butt.

THE MEDICAL MARIJUANA GROWERS GUIDE 1

Dedication 2
Disclaimer 8
The Medical Marijuana Growers Guide 9
INTRODUCTION 10
The Grow Room 15
Equipment List 19
Electricity 20
Lighting 24
ENVIRONMENT 28
Light 30
Air 32
CO2 34
Humidity 36
Water 38
pH Balance 40
Ambient Room Temperature 41
Soil / Grow Medium 42
Grow Pots 45
PLANT NUTRITION 47
Biostimulants 48
Primary Macronutrients 51
Secondary Macronutrients 52
Micronutrients 53
GROWING NO-TILL STYLE 54
Compost Tea 56
Fermented Tea 59

Topp Dressings *60*
Living Soil *61*
CEC *65*
Soil Composition *66*
Soil Mix *68*
PLANT DIAGNOSTICS *71*
Nutrient Deficiencies *73*
Nutrient Toxicity *74*
Mold Issues *75*
Pest Control *77*
Feeding Schedule *81*
Plant Behavior *82*
Five Stages Of Plant Development *84*
The Germination Stage *84*
The Seedling Stage *86*
The Vegetation Stage *87*
The Pre-Flowering Stage *88*
The Flowering Stage *89*
Ripening / Seed Set *90*
Growing From Seed *91*
Growing From Clones *93*
Growing For Recreation *98*
Growing Techniques *99*
Super Cropping *101*
Topping *102*
Lolly Popping *103*
Monster Cropping *104*

Scrogging	*105*
Thinning And Pruning	*106*
Topping Your Roots	*107*
Fimming	*108*
Agro Sonics	109
Yield	111
The Outdoor Grow	112
Choosing The Right Strain	114
Propagation	115
Harvest Time	116
Drying	*116*
Curing	*118*
Cannabis	120
CBD Rich Strains	*122*
Indica or Sativa	123
Benefits of Cannabis Indica	*125*
Benefits of Cannabis Sativa	*125*
Cannabis Root	127
Marijuana and Spirituality	129
Marijuana And Addiction	133
Medical Marijuana	136
How Cannabinoids and Neurotransmitters work	137
Top Ten Health Benefits	140
Raw Marijuana	143
Why raw?	*143*
Cannabis Extracts	144
Terpenoids	146
Flavonoids	146

Know your numbers 148
Summary 149
Bibliography 151
Net Work 152
YouTube Playlists: *152*
The End 153

Disclaimer

This book is a guide, created for educational purposes.

I am a licensed medical marijuana producer licensed by Health Canada under the MMAR program. The information in this book is in no way shape or form intended for illegal use.

The term medical marijuana is a term used by Health Canada in reference to what they identify as legal marijuana.

For serious illnesses, you should consult a doctor before using medical marijuana.

In some countries medical marijuana is illegal. I am not advising you to break the law.

Any attempt to grow or utilize medicinal marijuana must be done at your own risk.

I advise you to consult a doctor and acquire the appropriate licensing before you grow and use medical marijuana.

I am not responsible in any way shape or form for any loss or legal ramifications incurred by anyone while growing, using or distributing medicinal marijuana or from using any information in this book.

© 2016. Chef Derek Butt

The Medical Marijuana Growers Guide

Natures Pharmacy.

The Medical Marijuana Growers Guide starts out as a quick start guide on cannabis cultivation. Learn how to grow medical grade marijuana in small spaces or scale it up and feed the world.

The guide starts out with the basics then introduces advanced growing techniques that will increase quality and yield.

Biostimulants is the new frontier in plant nutrition. Learn how you can reduce your nutrients while increasing yield, quality and the plant's ability to protect itself from pests and diseases.

This guide will take you thru the different stages of plant growth and show you what is needed to achieve the plants full genetic potential.

The guide will also walk you thru the drying and curing process to ensure quality that any connoisseur can appreciate.

Plants are autotrophic, which means plants can produce their own food with only 17 essential elements. Carbon, hydrogen, oxygen and 14 other mineral elements that aid the plant's ability to harness energy from the sun, water thru the roots and carbon dioxide from the air, thru its leafs.

This food energy is stored as sugars and is used for plant metabolism, the production of fruit and flowers and the plant's defense system. This energy is also used in the root zone to feed microorganisms that feed the roots, which feed the plant and improve bioavailability.

INTRODUCTION

Cannabis Cultivation

A good grower is a good record keeper. You have to document everything you do. Grower's say the third grow is the first grow. The first grow you are getting to know the strain, the second grow you are tweaking your recipes and making small adjustments in your environment. The third grow you should know your strain well enough for the plant to reach its full genetic potential.

There are many how too grow books on the market today but I wanted to write something that reads more like a guide than a manual. I am pleased that I have this opportunity to share my many years of experience and knowledge that I have gained along the way.

The marijuana industry is much bigger then most people realize. The marijuana industry is bigger then all other industries put together, it is too big to centralize. This is why corporations and governments will not be able to monopolize and control the market.

The marijuana industry is a horizontal collective, meaning that everyone has equal opportunity to thrive. Marijuana has and always will be a cottage industry. A network of many growers that provide a rich diverse source of genetics. The mom and pop operation, the backbone of our economy.

People must be able to grow marijuana for themselves, as a vital food source and as a herb for its medicinal benefits. I believe marijuana should be scheduled appropriately under food, in the food and drug administration act before it truly becomes legal.

So long as cannabis is classified as a class one drug with its five thousand patents, it will always be heavily regulated. Cannabis as a narcotic should be regulated but cannabis doesn't become cannabis until you light it up or apply enough heat for the decarboxylation process to happen. Only then will marijuana become a narcotic.

Marijuana is essential in our diets as a complete food source as well as a medicine. Food is medicine. The cultivation of marijuana and its use as a medicine has been well documented for over eight thousand years. Ancient civilizations have evolved and thrived entirely around the marijuana plant. Weed is all we need.

The marijuana plant has commercial, industrial, medical, nutritional and recreational benefits. Marijuana can completely fuel the industrial revolution, provide us with everything we need and at the same time benefit the environment.

Marijuana is about to be legalized nation wide in Canada and when it does, we will see a shift in power. Industry as we know it will be downsized by seventy percent as the hemp/Marijuana industry takes hold of our thriving communities. As oil companies, pharmaceutical companies, alcohol companies, and law enforcement agencies loose their grip, we will be able to grow a more sustainable future. A future of peace and prosperity for all.

All we have to do is get those politicians high so they can gain a higher level of reasoning. People who are cannabinoid deficient do not think clearly at all. A cannabinoid deficiency can have a severe impact on one's neurology. Liberate the herb and watch us become one of the healthiest and wealthiest nations in the world.

I am afraid that legalization in Canada may just be another form of regulation. They want to take marijuana out of the hands of the criminal but pot farmers are not criminals. They have been criminalized by a corrupt injustice system. Pot farmers are more like saviors then criminals.

As far as keeping marijuana out of the hands of children, well we are capable of parenting our own children. The liquor control board has been unable to keep alcohol out of the hands of our youth, so I don't think they are going to be more effective with cannabis.

The College of physicians and surgeons forbids doctors to talk about medical marijuana, never mind prescribe it, yet there is a handful of doctors who appear to know nothing about medical marijuana, yet they are prescribing it at an alarming rate. We went from 2000 patients to forty thousand patients practically overnight. These doctors made millions by overwhelming Health Canada and essentially shutting down the MMAR program in the process. We are fighting this move in court, meanwhile, Health Canada is issuing LP licenses but you need a political connection to get one. Thousands have spent millions for an LP license with no hope of getting one. Perhaps after legalization, these LP applicants will be granted a license.

We won the court case, the decision is in. We will be able to grow our own medicine. I don't know whether to laugh or cry. It's not like I need permission to do the right thing.

People are facing life-threatening illnesses, people who really need this medication do not have legal access to it. I tried for eight years to get a prescription without any success until I decided to sign up with one of those so-called marijuana doctors for treating my arthritis.

These doctors will only prescribe marijuana as an anti-inflammatory. I did not lie to the doctor, I am using marijuana to treat my arthritis, but I am also using marijuana to treat a more serious condition. I almost did not get the prescription because I started out the interview talking about treating my bipolar disorder.

The medical industry will not embrace marijuana as a herb. They need to be able to prescribe precise dosages and that is not possible with a herb.

Pharmaceutical companies are working on cannabis-based medications as we speak and they are registering patents in an attempt to control the industry. They will not be able to patent a plant, but they can patent what the plant can be used for. These pharmaceutical companies are creating a monopoly on the market and there is very little concern for patient needs.

There is no good to be said on synthetic THC. Pharmaceutical companies that synthesize THC clearly do not know how THC works.

I no longer self-medicate, medical marijuana is now prescribed to me. I have a license to grow medical marijuana and for the first time in my life I can get up in the morning knowing that I have the right kind of medicine that is affordable. A medicine that is having a positive impact in my life, without any undesirable side effects, like incarceration and a criminal record. Looking after a garden is therapy in itself.

I will not buy commercially grown cannabis from an LP and I will not buy cannabis-based pharmaceuticals. Pharmaceuticals come with a seven percent fatality rate, popping pills is like playing Russian Roulette.

Fresh raw marijuana is four times more effective at restoring balance and has enabled me to regain some functionality in life without being high all the time.

Back in the day, we did not have YouTube or the wealth of resources we have when it comes to learning how to grow marijuana. We just planted seeds and learned thru trial and error and a little luck.

Mother Nature can be bountiful or devastating. Whether you are a hobbyist, a commercial grower, a medicinal grower or a recreational grower, this guide will help you gain an understanding on how to cultivate marijuana.

This guide will get you started and most of what you will learn will come from your own personal experience and the experience of others.

The Grow Room

Creating A Living Breathing Ecosystem

The way I see it, there are two ways of going about designing and building your grow room. You can start with a space with limited room and figure out how many lights you can use or how many plants you can grow or you can design your room around how many plants you are licensed to grow or how many plants you want to grow in the room you are designing and building. I think that was the longest sentence ever.

How big you build your room depends on how many lights you want to put in there and how many plants you want to grow. The optimum growing area for a 1000w light is 4-5sqft. Apart from the optimum growing area, you will need to keep your plants off the wall and out of corners, so account for 1 foot of space around your grow area. Also allow room for easy access for feeding, watering and maintaining your plants. Give your plants space, ideally, they should not be touching each other.

B1000w light bulbs burn at approximately 2300°F so if your room is too small you are basically building an oven. If the room is too large for your operating systems to keep up, your plants will choke.

Choose or build a room with high ceilings so you can create some kind of ecosystem that will support life. Leave yourself room to walk around the garden and work comfortably.

Your grow rooms should be sealed. No air or light leaks.

Your grow room has to be built so you can exhaust 100% of the air in the room within one minute. Because heat rises, you have to vent out from the ceiling.

Don't forget to install the appropriate size carbon filter on your exhaust line and remember that it will slow down the system.

You also have to replace that air and most of it can be sucked in thru passive convection but you will need a fan venting in on the ground level. All in all, the plants need room and air to breathe, so getting the numbers right is critical and I will give you the numbers as the guide progresses.

Hang your lights on chains about 4 or 5 feet apart and mount all ballasts and other heat sources above the plants, up in the rafters if you can.

You will probably have to install a serious air conditioner or you just might have to take the summer off, if you are an indoor grower.

You will also need to install oscillating fans to keep the air moving within your garden. Plants need fresh air and the exercise when bending in the wind. The movement damages the plant on a cellular level, which releases growth hormones, which increases growth and vigor.

Remove anything from the grow room that can harbor pathogens like carpets, furniture, curtains, pets, and Grandma. Clean and sterilize the room and keep it that way. A preventative maintenance plan works better than waiting until things get funky before you start cleaning up.

Paint the walls and ceilings a flat latex washable white to reflect the light in your grow room. Do not use that white plastic that you get at the hydroponic store for the walls and ceiling. In time, pathogens like mold will start growing behind the plastic.

I use 5ml plastic on the floor to catch runoff from the plants and it allows me to flush my plants without too much fuss. In effect, I am creating a non-recovery hydroponic bed where I can grow out of pots, filled with grow medium. Between each crop, I will lift up the plastic and make sure it's clean and dry under there.

You could also install a drain into the bed for those days when you have to flush your plants.

Try not to build stuff out of wood, wood harbors pathogens. Use metal shelves or metal tables.

You will need timers for your lights, fans, and ventilation, if not on a thermostat.

It might be a good idea to board up your windows or install metal bars on the ground floor for security reasons. Cover the windows with 5ml. white and black plastic to prevent any light leaks. You don't want a beam of light shooting off into the stratosphere. I will leave closed blinds in the window so everything looks normal from the street.

You will need a room for vegetation on an eighteen-hour light cycle and a room for flowering on a twelve-hour light cycle. You could get by with one room and adjust the light cycle to 12 hours when you want to trigger flowering. Things speed up pretty fast when you have a vegetation room that is constantly feeding one or two flowering rooms.

You will also need a room for cloning and propagating but it can be done in the veg room if there is room to work.

You will need a room for drying, curing, and storage. Preferably the room would be lined with cedar to prevent mold.

You will also need a quarantine room for introducing new clones from another garden (not recommended) but it's fast and convenient. When your plants get sick or infested they should be isolated.

Make sure you can lock your rooms.

You will need room for circulating fans. Leave a couple on 24/7 to keep the air moving at night, which prevents mold. I usually place my fans in the corners of the room since that is where mold usually begins, in stagnant corners. I have yet to buy a wall-mounted fan that actually works for more than a few months.

If I had the opportunity to build a room from scratch, it would be round and dome-shaped. I would also have to say that the bigger the dome is, the easier it is to create and maintain a living breathing ecosystem that is independent rather than dependent on the outside world.

All this is ideal but we all have to make do with what we have. You might have to start out growing in a closet or a little grow tent in the corner of a room. You can grow for next to nothing outdoors if you have a safe location. Learning to work with what you have is critical and as you grow, you will acquire the things you need along the way to up your game.

Equipment List

The Basics.

I will attempt to put together a list of equipment that you may or may not need. The list will probably be incomplete but it will help you as a guide.

High-Pressure Sodium ballasts (HPS) and light bulbs. Metal Halide light bulbs (MH) and ballasts, You can run red or blue lights on most new ballasts.

LED lights, (Light emitting diodes) and T5 fluorescent lights. UVB lights. (ultraviolet light)

Carbon filters, HEPA filters, exhaust fans, oscillating fans, timers, thermostats, hygrometers, a CO_2 system, heaters, air conditioners, dehumidifiers.

Did I miss anything? Pumps, hoses, clamps, nutrients, supplements, natural insecticides, grow medium, grow pots, water tanks, extension cords, gardening tools, a coffee machine, drying racks, seeds, clones and there was something else. I can't think of it right now. Do you get the picture? The list can go on and on. A sulfur burner and sulfur.

5ML plastic, white on one side, black on the other. Tuck Tape, flexible and fireproof ducting for ventilation. Coffee, zigzags, a bong, a 10X magnifying glass and or a microscope.

Protective clothing like a hat, sunglasses and rubber gloves. Shit man, there was something else. Oh yeah. Safe practical transportation, nothing too flashy.

Electricity

The Transformation.

I see a grow op as a living breathing organism with many systems that creates a synergy for life to thrive. Hydro is the blood running thru the veins of a grow op. Electricity transforms into light, which is harnessed by the plant thru photosynthesis.

The optimum level of light for growing is 40-50w per square feet of growing space. To determine how many lights you can use and how many plants you can grow in a room and determine some kind of a yield, you have to consider the available square footage of the room.

One of my rooms is 10'x20'=200sq.ft. I go on the premise that one 1000w light covers an area of 4'x4'. Right now I have a 4'x18' bed and a 4' x 8' bed totalling 104sq.ft of optimum growing area in the room. Most growers will choose 4x4 for optimum growing but if you are using multiple lights in a room then all the lights will be bleeding into each other and I believe you can stretch those optimum growing parameters to 5'x5'sq.ft. per light.

I have eight 1000w lights in the room that can be reduced to 600w if needed and I will grow anywhere from one to four plants under each light, expecting a yield of one pound per light. If you can accomplish that, you are considered a good grower in my book. Some growers claim to get two pounds per light or more. I must say though, it is not as simple as placing a few plants under a light. There is much to learn about growing techniques to get optimum yields.

It is a good idea to place all sources of heat like ballasts and ventilation fans outside of the grow room or above the growing area. What we are striving for is to be able to run at a 100% capacity while maintaining control over temperature, humidity, light, and air (CO_2 and oxygen) at optimum levels.

Do not try to grow more plants then what the indoor environment can sustain. Your yield will not go up, it will go down.

A 1000w MH or HPS light bulb will draw 10amps on a 120v circuit but it will only draw 5amps on a 240v circuit. A 600w light will draw 6amps on a 120v line and 3 amps on a 240v line. When using these lights, use a 240v line. If you are not an electrician then get one or consult with one. I know it is a breach of security but you have to know what you are doing when playing with electricity.

Do not exceed 80% of your capacity. Meaning, if you have a 100amp service, do not draw more then 80amps at any given time. You need the headroom. If you burn any hotter you will burn up. You will be running too hot and eventually, you will have a meltdown with disastrous consequences. Don't go there.

When calculating how many lights can go online, don't forget about everything else you will need to run your operation, like exhaust fans and air conditioners. 100watts will draw 1amp or .5amps on a 240v line.

Time to gather that list of equipment and figure out how much power you are going to need or how much equipment you are able to use. Know your numbers. I am not an electrician and I am smoking a joint right now, so take my advice with a grain of sea salt. There, my ass is covered. Failure to plan is a plan to fail.

Digital ballasts are the popular choice but they operate on much higher frequencies than the old school magnetic ballasts. Digital ballasts are hard on bulbs and they have to be replaced more frequently.

A digital ballasts EMF rating is a lot higher and there is much conflicting science on whether electromagnetic frequencies are good or bad for your health. In any event, your lines should be as short as possible. The line from the ballast to the bulb gives off the most EMF's. Ideally, it should not be more than six inches away from the bulb, so try to design your room with a longer AC line and a shorter ballast line. The shorter your lines, the cooler your system will run and you will use less power.

You can grow one plant under a light and train it, knowing that your vegetation cycle will be longer or you could grow four or eight plants under the same light acquiring the same yield but with a much shorter growth cycle. Time is of the essence but growing your plants larger is the only way to scale up if your license is limited to how many plants you can grow. There will be restrictions on how high or should I say how tall your plants are allowed to grow under legalization. The limit will be 100cm in Canada, so learn how to scrog your plants.

Generally speaking, a good grower can yield anywhere from one ounce per plant to a pound per plant or more growing indoors. Some growers can get up to ten pounds per plant growing outdoors.

All electrical utilities have to be mounted off the floor, preferably waist high and not near any water tanks, pumps or hoses. Whenever you work with water you have to be prepared and able to handle a flood without electrocuting yourself.

Play safe when working with electricity, wear shoes and don't stand in puddles. Make sure the power is off when you are wiring.

Lighting

Light is the most limiting factor in a grow room. Not enough light and your plants will not grow, too much light and your plants will stall. If you are serious about growing marijuana, you will need to invest in Metal Halide and HPS lights. They are the industry standard and if you want to up your game even further, those double-ended bulbs are more intense. A 600w double-ended bulb puts out as many lumens as a 1000w bulb. They have many advantages and are used by professional horticulturalists.

Double ended lights cost twice as much to buy but they are more intense, they cover a wider area. The color spectrum is more efficient and the par rating is 22% more. These lights will increase growth but only if you increase air ventilation and CO_2 uptake. With these lights, you will have to run your CO_2 at about 1500-1800 ppm to gain the benefit from intense lighting.

Care must be taken when using these lights. They burn very hot, the inside of that bulb is about 2300°F and they are very bright like the sun. Do not look directly into the light because you will eventually go blind. Always use protective sunglasses when working with these lights.

I use both MH and HPS lights in my grow rooms. Plants need both red and blue light in both vegetation and flowering cycles. You can buy super HPS bulbs that have adequate blue as well as red but they are expensive. MH gives ample blue light and the HPS bulbs give off ample red light, the two together work well.

HPS bulbs are very inefficient. Fifty percent of the energy is used for green and yellow light, which the plants do not need for growth.

I will supplement my light with UVB light between 285 and 350nm. This frequency range will increase the cannabinoid count in your herb, especially THC. THC is a natural sunblock so in an attempt to defend itself from ultraviolet rays your plants will increase their levels of THC by as much as five percent. Treat your plants with ultraviolet light in the last two weeks of flowering to get these results. These UVB lights have a limited range of about eight inches, so any buds beyond that reach will have significantly less THC. The double-ended bulbs give off ultraviolet frequencies so that's another plus for them.

Some of the new five band LED lights cover the ultraviolet spectrum, which growers use to produce more oils to make extracts.

If you use digital ballasts, you should replace your bulbs every two or three crops for optimum yields. LED lights will last five to ten thousand hours. Those double-ended bulbs will last for 10,000 hours and they will only lose 5% on efficiency.

In western cultures, we prefer 1000w lights for some reason. In Europe, 600w lights are preferred. I prefer to use 600w lights, go figure, I was born in England but I immigrated to Toronto, Canada in sixty-seven and now I live on the West Coast. 600w bulbs are more efficient and cost effective then 1000w bulbs.

1000w lights have to be placed three feet above the canopy and 600w lights have to be placed two feet above the canopy. These bulbs are intense and they run very hot. Placing your lights too close to the canopy will slow down growth, burn your plants, reduce cannabinoid content and we don't want that. Double ended bulbs should be placed around eight feet above the canopy.

If you grow your plants too close too the light, your herb may start to de-carboxylate from the heat that's displaced from the bulb. Once your herb is de-carboxylated it becomes psychoactive, a narcotic.

Some growers try to get the lights as low as possible to achieve tighter internode growth, which will result in bigger buds but you are playing on the edge, sacrificing quality for quantity.

With LED lights you can bring them as close to nine inches above the canopy.

Don't forget you are growing in a three-dimensional environment, so if your plants are three feet under the light you have one to two feet of optimum growing space to grow your canopy. This means you will have to train your plants to grow in this area. Instead of growing tall Christmas trees, you want to grow short bushes with many bud sites.

LED lights are becoming more popular and I would recommend them for the growth cycle and to supplement the flowering cycle. You can bring your lights close to the canopy and get very tight internode growth. LED lights do not need any special wiring. You will save 40% on your hydro costs and you have more control over the color spectrum. Some of the newer lights come with built-in digital timers.

They are still very expensive to purchase and you are looking at a big investment initially.

Some LED lights work well for both vegetation and flowering but most of the technology out there is out of date. You have to be careful with what you buy. Do not use LED's for cloning, they are too intense.

Newer LED lights also have white diodes for full spectrum light. I would recommend three-watt chips because they are more efficient. Some LED lights can be focused for deeper penetration into the canopy and can also be diffused to cover a wider area. Some LED lights are full spectrum while others focus on the red and blue light. Marijuana doesn't need green, yellow, UV or ultraviolet light to grow but a full spectrum will enhance growth.

The amazing thing about LED lights is that you can change the color spectrum and change the flavonoid and terpene profile of your herb. By fine-tuning light frequencies, you can change how your herb tastes and smells.

Even though MH and HPS are the industry standard, I would suggest taking a serious look at LED lighting before you make an investment.

You will also need fluorescent lights for growing seedlings and clones. They do not require intense light and you can keep the light close to the canopy for tight internode growth. You can keep your fluorescent lights about six inches above your plants. 4ft. HO T5 lights work well. Some growers use T5 lights in the veg cycle and they are getting good results. Some growers will take it a step further and replace the HO T5 bulbs with LED T5 bulbs for better performance.

With the technology that is available today, growers can formulate different light recipes that they can use at different stages of growth.

ENVIRONMENT

Environmental factors to be concerned with is air, water, heat, light, humidity, and nutrients.

The environment is everything. You can have the best seed stock in the world but if your environment doesn't provide your plants with what they need, the quality and yield will suffer. Everything around you, including yourself, is an environmental factor that has to be controlled.

The air we breathe the water we drink, the food we eat all play a roll in our health and the same goes for your plants.

If you are comfortable in your grow room then your plants will be comfortable and be able to thrive.
Plants need a certain amount of stress, which releases growth hormones and builds up defence mechanisms. In essence, you are playing God or should I say Mother Nature.

Marijuana grows well outside, the elements of nature provide a very harmonious environment for marijuana to thrive. It can grow just about anywhere, it's a weed but growing many plants in a contained environment requires some skill and finesse. Nurturing those cannabinoids to their fullest potential is an art form unto itself.

Prevention is key to a healthy environment, which is why I strongly suggest starting from seed. When you start from seed, you are off to a good clean start because the seed is clean with no pests or eggs, no pathogens, and no diseases. Starting with a clean sterilized environment and good seed stock with the desirable genetics is the ultimate way to go. However, starting from seed is a hit and miss situation. Even though a batch of seeds can come from the same parents, every seed is different. Seeds are expensive to buy and you are not going to get a consistent product so when you come across a really good plant that is better than the others, clone it. Every clone will be almost identical to the mother plant.

If you start from clones from another garden you could be inheriting all the problems form the garden that they came from and you will have a lot of work ahead of you to keep everything under control. Once your grow room is infested or infected with pathogens, you pretty much have to shut down, sterilize and start over again with fresh seed stock, if you want to operate completely pest free, so make sure those clones go into quarantine before you introduce them to your grow room. Some of the eggs that pests lay can remain dormant to ensure the survival of the species.

As mentioned, you need to scrub the air and place filters on your vents, this will keep out pests and help keep the air free of pathogens. Wear protective clothing in the grow room. Do not come in from another garden unless you change clothing and footwear. You might want to consider wearing gloves, a mask, and a hair net. You can't be too careful.

Light

Light is electromagnetic radiation. The color spectrum that plants use to grow ranges between 400 nm and 700 nm. 400 nm are smaller wavelengths that resonate into the color blue. Plants need blue light for vegetative growth. At the other end of the growth spectrum, the wavelengths are larger and resonate as red light at around 700 nm. Red light is utilized by the plant for root growth and flowering.

Light is the most limiting factor in a grow room. You need the right intensity, not enough light and your plants will not grow, too much light and your plants will stall. Fort to fifty watts per square foot will provide enough light for optimum plant growth. If you use high-intensity lights then other environmental factors must also be intense enough to process the light.

The quality of light meaning the color of light is also very important. Plants use red and blue light to grow but infrared light and ultraviolet light enhances plant growth. Ultraviolet light can be used in the last two weeks of flowering to stimulate resin production, resin is a natural sun blocker, which the plant produces to protect itself from ultraviolet radiation. Ultraviolet light can increase the quality of your medicine by as much as five percent.

Par is photosynthetic active radiation. It is how we measure light in the photosynthetic range of plants, which measures light between 400 and 700 nm.

Thanks to technology, growers can now formulate their own light recipes that change thru out the growing season. At the beginning of the growth cycle, you want blue light but you also need red light for root stimulation. Thru out the vegetative cycle, you want to decrease the red light and increase the amount of blue light. Once you get into flowering you want to decrease the blue light and increase the red light and as you get into the ripening stage, ultraviolet light will enhance your cannabinoid profile.

Light is also measured in Kelvins. 3000k are your warm tones, the red tones for flowering and 7500k are your cool tones, the blue tones for vegetative growth.

Plants convert light into sugars and we call this process photomorphogenesis. It is commonly known as photosynthesis. These sugars are stored in the plant as carbohydrates and starches and are used by the plant as energy for plant metabolism, for protection against pests and diseases and for flowering and fruit production.

The vegetation light cycle is 18 hours of light for the day and 6 hours of total darkness at night. Do not interrupt the light cycle, not even with a green light. The flowering light cycle is 12 hours of light for the day and 12 hours of total darkness for the night cycle. Again, do not interrupt the light cycle.

The light from a candle is one candle watt and that is enough to trigger your plants.

Air

Air is 78.09% nitrogen, 20.95% oxygen, 0.93% argon, 04% carbon dioxide plus trace amounts of other gases. Air also contains about 1.0% water vapor at sea level.

Air quality is critical for plant respiration. Plants have to breathe and they breathe thru their stomata, which are openings in the leaves. Plants breathe in carbon dioxide during the day and breathe out oxygen. During the night they breathe in oxygen and breathe out CO_2.

Air cannot be stagnant in your grow room. Air has to keep moving, adequate ventilation and circulation are critical for plant respiration. You must be able to recycle all the air in your room within one minute and you should use a carbon filter to scrub the air of dust and pathogens. Your carbon filter will also remove any odor that your plants are producing. You can take this one step further and install HEPA filters.

You will need to vent in from the ground level and exhaust out from above, heat rises. You need to be able to exhaust all the air in your room within one minute. If you are growing in a room that is 10x10 feet wide and 8 feet tall, that is equal to 800cu.ft. (10x10x8), you will need an exhaust fan that is rated at 800cfm. Meaning your exhaust fan will exhaust 800 cubic feet of air in one minute.

You have to install a carbon filter venting out that is large enough to handle the system. You must be aware that the filter will add a lot of drag to your system so go a little overkill on the exhaust. Don't go too overkill on the exhaust because the filter needs time to scrub the air as it passes through the carbon filter. The size of the carbon filter you choose depends on the CFM rating of your exhaust fan. Most people will vent in and out from outside which can lead to many problems. A professional grower will create an indoor ecosystem by recirculating the air in a closed system. One should also use HEPA filters on their exhaust fans.

If you vent fresh air in from the outside make sure you have a filter on the vent to prevent little pests from entering the grow room. Those little buggers hone in on your heat signature.

Do not vent directly outside or up in the attic. Vent down into the basement or a crawl space under the house so the heat can dissipate before venting outside. This will help keep you off the radar from night vision and heat sensors.

You will also need circulating fans in the grow room. It is important to keep the air moving in your grow room 24/7. This will simulate wind in the natural environment and give your plants exercise, which produces growth hormones. The wind will strengthen the stems of the plant, which you need for heavy bud production. They will also help keep the air fresh and clean, preventing the air from becoming stagnant.

CO2

If you are growing a large number of plants in a contained area you may have to supplement the air with CO2. Plants need about 350-400 parts per million (ppm) for healthy plant growth and our atmosphere provides us with that but in a contained grow op, increasing CO2 levels up to about 1200 ppm can increase your yield by as much as 30%. If you use these double-ended high-intensity lights you will need to supplement the air with about 1800 ppm of CO2 to gain the benefit of the intense light and to prevent your plants from stalling.

Smaller crops can get away without using a CO2 enrichment system. You may not be able to justify the cost or the risk of operating the system or having CO2 tanks delivered to your door once a week.

All in all, a CO2 system should increase your yield by 10-30% but that depends also on light intensity and many other environmental factors. Do the numbers and figure out what is best for your operation.

Add a few floor fans to keep the air in your room moving. CO2 is heavy and will drop to the ground.

When using a CO2 system, you want to recirculate the air in your grow room rather than just vent in and out. In essence, you will be creating an indoor ecosystem. You do not want to waste your CO2 by venting it out as fast as you are venting it in.

Be careful with this stuff, CO2 can be very toxic for us, make sure your system can be regulated accurately with the appropriate equipment.

Humidity

As the plant transpires gases, humidity builds up under the leaves, provide adequate air circulation and make sure your plants are pruned to encourage air circulation. Humidity must remain consistent. Any sudden changes will add stress to the plant and make it vulnerable to disease and infestations.

For seedlings and cuttings, you want to maintain high humidity levels, your plants do not have a root ball to absorb moister. Most of the moister will be absorbed thru the leaves and you must maintain high humidity levels above 80%.

As your plants mature in the vegetative cycle, you can slowly reduce the humidity, perhaps by 5% a week until you get to about 60% for the growth cycle. Maintain 50-60% humidity throughout the growth cycle. It is important that your humidity levels remain consistent for optimum yields.

As you enter the flowering cycle, gradually reduce the humidity to about 40%. This will prevent mold and bud rot from invading your plants.

When drying your plants you also want low humidity levels, around thirty-five to fifty percent. You can regulate how long it takes to dry your herb by adjusting humidity and heat levels. The warmer and dryer the room, the faster your buds will dry.

You want to dry them slowly over a period of two weeks but some growers will take as little as three days to dry their herb.

Too much humidity in the grow room will inhibit respiration, which will slow growth and reduce yields.

Water

Water provides minerals needed for plant growth as well as hydrogen and oxygen. The quality of your water will have a direct reflection on the quality of your product.

Pure water is two parts hydrogen and one part oxygen. Pure water does not conduct electricity but the minerals in your tap water do conduct electricity. The more minerals that are in your water, the more conductive the water becomes. Mineral content can be measured by its electrical conductivity. Meters can be purchased that measure mineral content in parts per million (ppm).

I prefer to use tap water rather than reverse osmosis water. Filtering out the minerals and adding them back in is expensive and not necessary if your water is not contaminated with toxins like fluoride.

The chlorine in tap water is a mineral that plants need for growth. Have your water tested so you know what is in it and how to treat it. If there is fluoride in your water, it should be filtered out.

Groundwater that has been filtered thru the earth or well water can also be used but the condition of the water can change drastically as the weather changes. Keep a close eye on the pH and the mineral content of your water supply because it will not be consistent.

The important thing about your water is that it is clean and you have a consistent supply.

Keep your water cool. The optimum water temperature is 64-68°F. The warmer the water the less oxygen it has. Running a pump in your water and circulating the water will oxygenate the water. Keep that water moving and flowing like a river. Do not let the water sit for too long.

If you are worried about chlorine levels, 70% of the chlorine will dissipate in 30 minutes and it will take about 24 hours for all of the chlorine to dissipate.

Do not keep your water tanks in your grow room. It will be difficult to control humidity.

pH Balance

The pH (potential hydrogen) is the measure of acid and alkali levels. Seven is neutral on a scale of 0-14. Marijuana grows best in a slightly acidic environment, the sweet spot for growing marijuana is 5.5-6.5. You will need to check the pH levels constantly since everything that happens in your environment effects pH levels. Get yourself a pH meter, you will need to take accurate pH readings because your pH levels will determine how well and what your plant is able to process.

If you can't afford the meter, some pH tape will work if you are not color-blind. Just dip the tape into your solution, the tape will change color, match the color to the chart that is provided to get your pH balance.

Your plants get the final say on the ideal pH levels. You will probably be growing just females for bud production and like most females, they like to get the last word in.

Your plants will tell you what they need and your job is to observe, document and provide. Learning how to diagnose and treat symptoms is critical to the success of your garden. The color of the leaves, the stance of the plant, any spots or markings are all indications of what the plant needs.

When cloning, keep your pH levels at 5.5 for fast root development.

Ambient Room Temperature

The ideal ambient temperature for your grow room is 75°F. Anywhere from about 73°F-80°F is good but the closer to 80°F you get the more problems you will have with pathogens and pests. 80°F is optimum for plant growth but it is also the optimum growing temperature for everything else under the sun.

To maintain these temperatures you will need an air conditioner or a water cooler. Personally, I would rather take the summer months off or grow outside.

Marijuana plants can tolerate temperatures up to 90°F. but growth slows down considerably.

You will need a thermometer that also reads nighttime temperature (low temp.) since the temperature of the room must not drop more than 10°F at night. If so your plants will go into shock and internode growth will stretch.

Soil / Grow Medium

Your grow medium should be pretty close to 80°F for optimum growth. You want the medium warm enough to encourage microbial growth.

There are a few mediums to choose from. coco coir, rock wool or good old sphagnum and perlite are popular choices. You could consider aeroponics or aquaponics, utilizing air and water, if you were considering harvesting the roots.

I use sphagnum and perlite, making sure there is enough perlite for aeration. I do not add amendments to the grow medium because it gives me more control thru out the growing season.

There are a few different grow mediums to choose from and what you choose depends on what system you choose to grow in. If you plan on using a hydroponic system then you will be using rock wool. If you choose to grow in a non-recovery hydroponic system you will most likely go with sphagnum and perlite or coco coir. If you grow with organic nutrients, go with sphagnum and perlite for its absorbing properties. When growing organically you are conditioning the grow medium and the grow medium will feed your roots and your roots will feed your plants.

If you plan on using salt-based nutrients you should use coco coir because it is not as absorbent and will not retain as much salt in your medium. When feeding your plants with salt based nutrients your plants feed directly from your nutrient solution rather than the grow medium. Coco coir is easier to flush and keep clean. Coco coir is more expensive to buy but it can be re-used as many as five times if you remove the roots. You don't really need to add perlite even though some people do.

Most people use a mix of sphagnum and perlite. It is very convenient to buy your medium premixed and it will come with some amendments like dolomite lime to neutralize pH levels. It usually also has a wetting agent that helps water to penetrate the grow medium.

Water molecules are large and will follow a path of least resistance, meaning the water will stream down your pot rather than spread out and saturate the grow medium. Without a wetting agent like soap or Yukka, dry areas will develop and the roots will not grow in those areas.

The perlite will help aerate the grow medium and it will also help retain water.

When preparing a mix, some growers will add nutrients and fertilizers and each grower has their own unique recipe that they become accustomed to working with.

I prefer not to add fertilizers because I want full control over the different stages of plant development. Nutrient requirements differ over the different stages of development and it is also much easier to flush everything out when the time comes.

I fabricated my own mix but I am now licensed to grow 122 plants and I do not have space indoors to work with it, so I pay for the convenience of a premix and I do not add anything else to it.

If you mix up your own, add plenty of perlite for aeration and drainage in the root zone. The volume of perlite should be equal or less than the volume of sphagnum to ensure proper aeration and drainage.

The pH of sphagnum is 4.5 without the dolomite lime but that can be compensated for when mixing your nutrient solutions with pH up or down. You could use baking soda to raise the pH or lemon or vinegar to lower the pH.

Do not mix pH up and down together. A chemical reaction will happen that will melt plastic in seconds. If you add too much pH up to your mix and you add pH down it will burn your plants from the inside out.

Grow Pots

A plant will only grow as wide as its roots can spread. Picking the right pot size and shape is important. It depends mostly on how wide and how tall you want your plants to grow. Plants will grow tall and narrow in a tall pot and plants will grow wide and bushy in a wide pot. Thanks to gravity, wider pots hold more water then tall pots, so I always go for the wide pots for indoor and outdoor gardening.

The pot itself should be able to breathe, allowing air to penetrate into the roots. Fabric pots are ideal for this. Pots need to provide roots with air as well as contain water, so make sure you have adequate vent holes in your pots.

Fabric pots will dry the tips of the roots as they reach the outside of the pot, which prevents the plant from becoming root bound. This method will also invigorate new growth in the root ball, which will increase yield. You can also achieve this effect by drilling small holes in your pots. You will need to water more often as your plants grow bigger and faster.

Start your cuttings or seedlings in small four-inch pots and let the plant get established in the pot. The plant will establish itself underground first before it goes into full vegetation.

Transplant your plants into a medium-sized pot to finish of the vegetation cycle and then transplant to a larger pot when the plant is ready for flowering. This method will ensure optimum root growth and bud development.

I start my clones off in a 4" pot, wait until the plant gets established and then transplant into a 2 gallon pot. I will veg the plant to the pre-flowering stage and transplant them into 5gl pots for the flowering stage.

If your plants become root bound, growth will slow down because the roots are growing around the inside of the pot, looking for food and water. If you cut off the excess roots, this will invigorate new growth inside the grow medium, where the plant can drink and feed in comfort.

Instead of spending the money on drip trays, I just line the floor in 5ml plastic. It acts as a bed and you could install a drain for the runoff. I have recently purchased a few 10gl wash tubs. I will place a plant in the tub and flush the plant until the runoff is clean.

Remember, you should always water your plants with 5% runoff and you will have to flush your plants from time to time, having adequate drainage or a runoff system is high up there on the priority list.

If you are not growing a lot of plants, a bathtub or some kind of tank will work to catch runoff.

PLANT NUTRITION

I produce medical marijuana so I grow organically. A number of companies make organic solutions that work well. Some are not certified organic but they are a step in the right direction. They are convenient to use.

I have used General Organics, which is not certified organic but companies like Fox Farm produce certified organic solutions. They are expensive but the finished product is cleaner. There are many advantages to growing organic and the ideal solution is to grow organic soil, rich with microorganisms that produce nutrients for your plants rather then just feed your plants nutrients.

Plants are autotrophic meaning they can feed themselves from the nutrients that are provided by microorganisms in your grow mix/soil. Plants will be able to feed of bottled petroleum based nutrients but they will not be healthy enough to defend themselfs from the environment so more chemicals are used to protect the plant from infestations and unwanted pathogens. It is a downward cycle spraying chemical after chemical to keep your plants alive from being raised on chemicals.

The new way of thinking is totally old school. Instead of feeding your plants you feed and cultivate microorganisms in your soil, which provide the nutrients, the minerals, the enzymes, the vitamins and most of everything else your plant needs to thrive. There will be no need to feed your plants or spray them with chemicals.

Biostimulants

Biostimulants is the new frontier in plant nutrition, or should I say the old frontier. Commercial agriculture kills the rhizome layer, which is where microorganisms live. Microorganisms feed our plants but with that gone, the industry has to start pumping salt based nutrients into the soil to feed the plants. They use harmful pesticides and fungicides to protect the plant because the plant is no longer healthy enough to protect itself.

Agriculture has taken over what the plant can do for itself. As a result, plants will suffer from a low sugar content and will not taste as good as organically grown food. It will also lack in nutritional value and its medicinal properties.

Growing with biostimulants will increase quality and yield by as much as 20%. Biostimulants aid in the uptake of minerals and water. They will increase the Brix content, which is how we measure sugar content. These sugars are stored as energy in the leaves and are used to increase chemical reactions and plant metabolism. The goal is to achieve at least a 12% sugar content. If you can achieve this, the nutritional and medicinal qualities of your herb will improve.

Pests will not identify the plant as a source of food when the Brix content is high so it acts as pest control. This energy is also used to build the plants immune system and fight diseases.

This energy is also stored and used for flowering and fruiting. Plus this energy is also used in the root zone to feed microorganisms that feed the plant.

When you start feeding your plants bio stimulants it increases the plant's uptake efficiency. This means you can feed up to 50% less while increasing yield and quality.

Plants are autotrophic, meaning they can feed themselves. All they need is 17 essential mineral elements.

Photosynthesis is a process that plants use to synthesize nutrients from carbon dioxide and water, which is hydrogen and oxygen. The plant will harness the suns energy and store it as sugars, carbohydrates, and starches. This energy is used to draw up water from the roots and breath in carbon dioxide thru the stomata's located on the leaves.

They will breath out oxygen and reverse the process at night. This energy is used for plant metabolism and defense against pests and disease. The plant also uses this energy to feed rhizobacteria in the root zone. These microbes will make bioactive molecules called biostimulants. Biostimulants produce pectin, which lines the cell wall, which prevents mold from establishing itself on the plant. Pectin is the glue that cements cell walls together.

Biostimulants are enzymes that break down proteins into amino acids. Roots feed the plant, microbes feed the roots, which activates the genes, which tell the plant what to do.

Humic acid and fulvic acid are a good source of amino acids. It activates enzymes, buffers pH, increases voltage potential in the cell membrane, is a source of trace elements, improves flavor, nutritional and medicinal properties.

Seaweed extract provides growth hormones and trace elements. It provides B vitamins, which aid in stress. it activates enzymes and increases chemical reactions.

During the last two weeks of flowering, the plant will stop sending sugars down to the roots and the roots will shut down. To prevent this and keep the plant photosynthetically active, feed the roots molasses during the last two weeks.

Use 5 parts humic acid to 2 parts seaweed extract to make a foliar spray. Use it to fight stress by keeping the plant on high alert and fast reacting.

Primary Macronutrients

The three primary macronutrients are Nitrogen (N) for plant growth, Phosphorus (P) for flowering and potassium (K) as a regulator for overall plant health and reproduction. Nitrogen is the first number, phosphorus is the second number and potassium is the third number on your nutrient bottle, package or container.

Nitrogen stimulates vegetative growth. Use ammonium nitrogen in the beginning of vegetative growth. Cell growth will be smaller but cell walls will be stronger. Later in vegetation switch to nitrate nitrogen, this will grow larger cells but with thinner cell walls. Go easy on the nitrogen, especially the nitrate nitrogen because 30% of the plant's energy will be used to assimilate the nitrogen.

Phosphorus stimulates root growth and early flowering. 23 isotopes of phosphorus are known and some of them are radioactive, they provide energy to grow a massive root system and in turn, the root system will feed your plant.

Potassium is the health element during vegetative growth and flowering. During the vegetation cycle, you will need 1 1/2 times more potassium than nitrogen. During flowering, you will need 2 times more potassium than nitrogen. The plant will consume a lot of potassium, it is needed to assimilate phosphorous for flowering and fruiting (reproduction). Adding heavy amounts of phosphorous during mid-bloom is a misconception. At this point, you need to reduce phosphorous and increase potassium.

Secondary Macronutrients

The three secondary macronutrients that you will need is calcium (Ca), sulfur (S), magnesium (Mg). These nutrients may be in an all-purpose formula but they may have to be added to your recipe.

Magnesium aids in the production of chlorophyll, which is the plant's hemoglobin. Sulfur triggers flowering and calcium in the form of calcium pectate binds plant cells together.

Micronutrients

The micronutrients or trace minerals that you will need for healthy plant growth is boron (B), chlorine (CI), manganese (Mn), iron (Fe), zinc (Zn), copper (Cu). molybdenum (Mo), and nickel (Ni).

These nutrients are primarily absorbed by fine root hairs and the transpiration of gases. Allot of these minerals are in the soil if you plan on using soil and allot of these minerals are in the water that we drink. It is important to get an analysis of your water so you know how to treat it.

Some growers will use RO water, which removes all trace minerals and the guesswork in coming up with the perfect recipe.

If you are growing on this level you will probably be using Advanced Nutrients, which are very expensive and you really have to be an expert to use them.

Trace elements activate enzymes that produce a chemical reaction that stimulates metabolism and genealogy. They increase the uptake of secondary macronutrients like calcium, magnesium, and sulphur.

GROWING NO-TILL STYLE

Chemical fertilizers are mostly derived from petroleum, they are inorganic and can be absorbed by the roots of plants, however, they are pollutants, which can cause a die off and a population change of soil microbes. Built up unused residues which run into the water table create harmful tissue changes in the plants, which people consume as food and medicine. In addition, The use of chemical fertilizers promotes the incidence of plant pathogens like powdery mildew, Erwinia, Fusarium, Pythium, etc. The grower can end up in a vicious spiraling downward fall as they use one chemical after another to control the effects brought on by the others.

The objective of growing no-till style is to eliminate the need for harmful chemicals. The benefit is a healthier, more productive plant that doesn't require feeding or spraying toxic chemicals to keep the plant alive.

Growing no-till style indoors requires you to grow in larger pots, preferable fabric pots so your soil and the plant's root system can breathe. Basically, what you want to do is recreate the forest floor in your pot. By adding compost, worm castings and other amendments to build up the soils micro-biome, which is a community of microorganisms that will produce nutrients, minerals, vitamins and enzymes that will break down proteins into a soluble form. In essence, the microorganisms will be feeding your plants. Your job is to create a micro-biome and feed the microorganisms.

When your plants are ready for harvest you simply cut the plant down, pull out the plant's main root system and plant a new one in its place. No need to remove the roots. If your pot is big enough and you have the enzymes to break down the old root system, they will be converted into soluble nutrients that the plant will feed off. In time the condition of the soil improves and becomes more productive.

Grow a cover crop like alfalfa or clover, which is a nitrogen fixator. Meaning it will consume nitrogen in the air and deposit it in the soil.

Use a top dressing periodically and keep the soil covered in a leaf mulch. This will retain moisture, warmth, and air in the soil. (the rezone layer.)

Use compost teas to feed your soil and to foiler feed your plants as necessary. There may be times when you need a bacteria or a fungal dominant tea to treat unwanted pathogens or to correct an imbalance of microbes but most of the time a simple balanced tea will service.

Take it to the next level and start making and adding fermented teas to the mix.

You can also sprout organic seeds like alfalfa or popcorn seeds. When they have germinated, juice them in a blender, strain and add that to your mix.

Some growers will use coconut water for their powerful enzymes.

Compost Tea

There are many recipes available for you to use and there are may compost teas on the market that you can purchase. They all work to some degree and the advice is to use them all to achieve biodiversity. That can get expensive and there are times when you may want bacteria dominated tea or a fungal dominant tea but 98% of the time you will need a well-balanced tea. The big reason for this is because the plant will choose what it needs by feeding the microorganisms necessary to complete the nutrient cycle, while other microorganisms will lay dormant until the plant needs them. In essence, the plant chooses for itself what it needs. So providing a balanced tea will ensure the plant will have what it needs when it needs it.

Aerobic bacteria need oxygen to thrive so If you plan on brewing large batches of tea you will need a brewer that can incorporate oxygen between 6-9 ppm. You also need to agitate the compost. This will release the microbes attached to the plant matter. If you plan on making smaller batches for your home grow you will need a bucket with an aquarium pump. Get something that can at least clean a 60-100 gallon tank. If you limit your batches to 3 or 4 gallons it should work fine.

Brew the tea at 65-75°F. Within 12 hours you will have a bacteria dominant tea. Take it to 24 hours and the protozoa starts feeding on the bacteria and fungi begin to thrive. Fungi grow through the soil and serve to, bind soil aggregates together, help retain moisture, store certain nutrients, provide a source of food to certain other microbes, provide pathways for nutrient and moisture delivery, decompose organic material and displace disease-causing fungi.

The protozoa will multiply and feed off the bacteria providing soluble food for the plant when the plant needs it. This completes the nutrient cycle or life cycle. In thirty-six hours you should be able to brew a well-balanced tea. I would not take it past 36 hours.

I would not use a tea bag unless you put the bubbler in the tea bag. If you don't use a tea bag a sump pump should work for feeding or you can water by hand. You will have to filter the tea if you use it as a foiler spray and you should foiler feed often or when needed.

You can tell when the tea is done by the way it smells. It should smell earthy, like mushrooms. If it turns anaerobic it will start to smell like a sewer.

Diluting the tea with water will help spread those microbes around. Dilute one to one to feed your soil or one to three as a foiler feeder.

The best compost tea recipe in the world is very basic and simple.

5 gallons of dechlorinated water.

2 cups of worm castings.

1/2 cup of un-sulfured molasses.

Place these ingredients in a pale and brew for 36 hours. This recipe will provide you with the three groups of microbes to complete the nutrient cycle. Bacteria, protozoa, and fungi. The idea is to grow as many different strains within each group to achieve biodiversity. The problem with adding anything else to the recipe is you run the risk of upsetting the balance and the culture in your brew may crash. My advice is to keep it simple. Do not add anything else to the tea once it is done and does not add any humid acid to your tea. The humic acid will inhibit microbial growth in your tea. It is ok to use it in your soil.

Fermented Tea

Gather some weeds that do well and grow locally. They provide enzymes that thrive in your area. Chop the weeds up. About two cups of plant material to one cup of organic granulated sugar. Need the two ingredients together. The crystalline structure of the super will draw out the moisture in the plant. Place ingredients in a glass jar and cover the material with more sugar until it is completely covered. This will prevent oxygen from getting down into the plant material. Cover the jar with breathable clothe like a paper towel to prevent any bugs from getting in. Let it sit in a warm dark place for a few days or until the fermentation process has begun. Pour of the plant juice and use it to feed your soil or as a foiler feeder.

Topp Dressings

Here is where you can get a little creative but remember, the amendments that you are adding is food for the microbes not food for your plants.

5 gallons of pro mix or sphagnum and perlite with a wetting agent like yucca.

2 cups of worm castings.

2 cups of mushroom compost.

1 cup alfalfa meal.

1 cup kelp meal.

1 cup of neem meal.

1 cup of glacial rock dust.

1 cup of bone meal.

Dolomite lime.

You can play around with the recipe, just don't overfeed those microbes. Apply as a top dressing and water in your plants.

Living Soil

A living soil is comprised of a large variety of creatures, mostly microscopic and single-celled. Part of this life is the plant itself but billions of life forms which support this plant and microcosm are arranged hierarchically at a level in the soil to which they have evolved for optimum survival and the holistic function in their universe.

There are multiple interfaces in the soil. There are millions of small pores throughout, millions of various particles interfacing as aggregate; sand, clay, silt, rock, organic matter, humus and thousands or millions of roots interfacing these.

Besides these areas of contact or buffer, there are some broader distinct fields of transpiration between life forms which thrive within certain environmental conditions. This is why, as horticulturists, we may achieve living soil through minimal soil disturbance or no-till.

To describe these fields, first, lets talk about the soil's surface. Soil scientists call this the Detritusphere, not a very complex name when you consider what detritus encompasses. So here is where stuff falls; everything from leaves to poop and this is where the greatest velocity and frequency of decomposition occurs. The detritus is principally carbon-based. The elements of oxygen, nitrogen, light, and moisture combine with the microorganisms evolved in this environment to do their job of degradation through consumption. These organisms are specialized to use the components and fuel available in the top layer of the soil, let's say the top one to three inches dependent on soil type.

At a lower depth, they would not function similarly because the fuel would be lacking. The material processed as waste by these microbes is then passed down to the next set of microorganisms evolved to process that modified substance.

If the raw detritus is worked into the soil, without first being degraded by surface dwellers, then the subsurface microbes can become overwhelmed with the task and can easily use up any and all nitrogen at hand decomposing this organic matter, thereby depriving local plants of this nitrogen. This can result in what some refer to as nitrogen lockout or lock up.

The next interface is where openings are created by earthworms, nematodes and other larger creatures, called the drilosphere by scientists. This is an area where some of the previously described material is conveyed by the bugs and worms along with bug and worm poo and bio-slime. The bio slime created is important for binding particles and contributing to aggregation. Obviously, these create unique passageways for certain sized organisms, air, and water.

Branching off of these passages and stretching into the entire area which we call our living soil is a myriad of various sized openings and caverns. This area is referred to as the porosphere. This is where the meat and potatoes of the soil grow, is stored and is hunted. It is this zone which interfaces with the roots, which is called the rhizosphere.

Of critical importance is the conjoining matter, the particles which comprise the soil itself. These pieces once bound together by bacterial and fungal bio slime is referred to as aggregated material and how they choose is what forms the aggregatusphere. The aggregation is bound by fungal hyphae, roots and various gel-like polymers and carbohydrates excreted from plants and creatures alike.

When the gardener/horticulturist first mixes their soil, they can have some pretty good control over the size of pores created, balanced with decomposed organic matter.

The variously sized particulate creates the multitudinous openings and caverns which make survival habitats for certain small organisms like bacteria and archaea and hunting grounds and habitat for some larger organisms like protozoa, nematodes, and rotifers. These spaces flow with water and air allowing bacteria, archaea, and fungi to mine the stored/sequestered nutrients, from vermicompost (worm castings), compost, humus, clay/rock and other organic matter, which are then passed via the rhizosphere in a number of ways to the roots. There are miniature pockets of water bound to soil particles which are necessary to the survival of many microorganisms.

Methods of Nutrient Assimilation in the Rhizosphere

There are a variety of ways in which plants uptake nutrients organically/naturally. The majority of relevant current research indicates that most nutrients are derived from the predation of bacteria and archaea by protozoa and nematodes. The waste produced by the larger organisms is in ionic form, being directly taken up by the roots. In addition to this, there are mycorrhizal associations between certain types of fungi and roots whereby the fungi provide the roots with nutrients and receive nutrients in exchange.

The most active protozoa contributing to this nutrient loop are flagellates and naked amoebae, however, ciliates and testate amoebae cycle nutrients to a lesser degree in an anaerobic soil. As the flagellates and naked amoebae consume bacteria/archaea they utilize somewhere from 10 to 40% of the energy intake for sustenance, dependent on species. The excess is excreted in an (ionic) form directly available to the roots of the plants. This means a plant can receive a whopping 60 to 90% nutrient bonus from this exchange.

As I have indicated previously the plant is not necessarily passive in this process. Studies show that plants emit certain carbons from their roots which attract and feed specific types of bacteria/archaea. Once these bacteria/archaea begin to divide, they begin to thrive on the adjacent organic matter (using organic acids) and the population explodes, thereby stimulating a resultant protozoa population explosion.

We should not leave the bacterial feeding nematode out of this. They also cycle nutrients via the microbial nutrient loop in a similar fashion by predation of bacteria/archaea and excreting bio-available nutrients. One difference is that they require about 50 to 70% of the energy intake for sustenance, however, they are much, much larger. I suppose that due to their size, they cannot get to some spots that protozoa do. The other consideration is that bacteria can multiply every 20 minutes and protozoa every 2 hours, while nematode eggs take 4 to 7 days to 'hatch'.

Roots also exude various organic acids like carbonic acid, citric acid, malate, oxalate and several others. These acids solubilize sequestered nutrients into an ionic form which they can assimilate. [e.g. dissolved organic nitrogen (DON); phosphorus; (DOP)] Some bacteria and archaea (besides the nutrient loop previously described) excrete similar acids which degrade organic matter and provide nutrients directly to the roots or the soil solution (an area in the rhizosphere where nutrients are in solution) and some fix atmospheric nitrogen and are symbiotic with legumes.

Note: fungi also excrete similar organic acids to release/degrade nutrients from organic matter.

CEC

Where does CEC (cation exchange capacity) come into this picture? The CEC is your soil's capacity to hold nutrients. It is based on your soil components having a negative charge and holding on to positively charged nutrients. Various types of clay like bentonite, organic matter, and sphagnum peat moss have excellent CEC.

It is this researcher/gardener's understanding or hypothesis that the nutrients which are held in place in the soil are released by the various types of acids (citric, carbonic...others) mentioned previously. These acids are exuded by bacteria, archaea or roots to create hydrogen ions which then displace (exchange for) into the soil solution, the nutrient ions required by the plant. In the case of bacteria/archaea which have consumed these nutrients, they are themselves consumed by protozoa and nematodes which they expel as waste in ionic form nutrient immediately available to the plant, as previously described.

It appears that this method of uptaking the desired nutrient is more 'economically' viable for the plant. Rather than expending its precious resources to mineralize (release) these nutrients, the bacteria, archaea, protozoa, and nematode pull it off for her.

Soil Composition

The number one method of nutrient uptake listed above that the horticulturist can influence is the predation of bacteria/archaea by protozoa and perhaps nematodes. By ensuring a good soil base with a variety of pore sizes but with lots of adequate drainage, moisture retaining substance and composted organic matter, one will provide good habitat and hiding spots for these organisms to flourish.

When creating your soil mix bear in mind that you wish to create long-lasting spaces or pores of various sizes so it is best to include some very slow to decompose organic matter and some rock or sand-like particles along with some of your faster-degrading compost to see you through your first season as your soil matrix comes to life.

There is another sphere of influence in the soil which I feel is of importance and that is the interface between stone/rock and the upper portions of the soil. For container growing there is going to be variance in accord with your container size and depth and the way, you wish to arrange things. I do believe that there are groups of microorganisms (bacteria/archaea & fungi) which work at certain depths with limited to no oxygen which mineralizes nutrients from stone, rock and rock powders. In similar fashion to the surface dwellers, the nutrient waste which they process is passed up the chain and then to the roots. Within this hypothesis, there may be some logic in placing a layer of small stones or gravel in the bottom of a container. Of course this makes more sense in a larger, deeper container.

Anecdotally, I surmise that a variety of colors of rock/stone is beneficial. This is derived from the idea that as people we assimilate more vitamins and minerals by choosing diversely colored foods.

I hope I have conveyed that allowing microbes to live and function hierarchically at their optimum position undisturbed is how a horticulturist best achieves living soil. By leaving soil undisturbed fungal hyphae circuitry remains established, mycorrhizal colonization of roots takes place more quickly, networks of microbial nutrient exchange stay in optimum position.

Of course, it is a decision which each grower must make on their own, balancing what is feasible and convenient to the space available and to their lifestyle and ability. I can attest that my experience with this method of container growing is that the soil just seems to get better with each season.

It is important to keep it alive through additions of organic matter, topdressed and I believe a minimum volume of 5 gallons and 14 inches depth is important. A larger volume is likely better. Allowing the soil to be populated by small arthropods, nematodes and perhaps earthworms is of great value.

Soil Mix

It seems like every grower has their own unique recipe that works for them. Some recipes can get extensive with 15-20 amendments.

Diversity is the key, having as many different species of the three categories of microbes bacteria, protozoa, and fungi will complete the nutrient cycle, the cycle of life. Your main source of these three microbes will come primarily from worm castings. Your next main source of microbes will come from compost. Everything else that you add to the mix is a source of food for your microbes. Once you achieve a balanced soil your plants will be able to choose what nutrients it needs when the plant needs it by feeding the necessary microbes.

INGREDIENTS:

SOIL BASE:

Peat moss

Coco Coir

Perlite

Yucca

AMENDMENTS:

Worm castings

Mushroom compost

kelp meal

Alfalfa meal

Neem meal

Bone meal

Glacial rock dust

Dolomite lime

Fishbone meal

Sulphate of potash (rocket fuel for flowering)

Humic acid

There are many more amendments that you can add to the soil. You can go with manures and bat guano. Blood meal, fish meal, feather meal, basalt, green sand, oyster shell flour, gypsum, rock phosphate.

The thing is, a simple recipe will work and improve in time. Gathering all these materials can get expensive but when in place you will have created a soil that is alive, a soil the grows and improves with age. You can incorporate different amendments over time with different top dressings.

When transitioning from vegetation to flowering the plant will know what to do. The shorter days and longer nights will tell the plant to start flowering and the plant's root system will start feeding the necessary microbes who in turn will feed the plant what it needs for flowering.

The plant will start feeding different microorganisms that produce different nutrients the plant needs during flowering. What we can do to help that along is keep the plant slightly acidic during vegetation. (6.2 pH). This will make micronutrients more available when needed.

Keep the soil more alkaline which will make macronutrients more available when needed. (6.8pH).

Your soil is not an expense anymore. Now it is an investment that keeps growing.

PLANT DIAGNOSTICS

When diagnosing plants the first step is to establish what kind of plant you are treating and what they are prone too. This helps as a guide because many people have experienced the same problems that you are having.

Marijuana is susceptible to powdery mildew, mold, spider mites, fungus gnats plus a host of other pests and pathogens.

The next step is to establish if the problem is environmental or pathogenic. If the problem seems to happen overnight it is environmental, if the problem happens over time like a slow onset then the problem is pathogenic.

If the problem is environmental it can show up as a nutrient deficiency or a nutrient toxicity. It could be anything so go thru your systems and use the process of elimination to get to the root of the problem. Determine if the problem is an infestation or if it is too much or not enough light, water, or CO_2.

If you are experiencing a pathogenic problem then you are entering a world that is built on many layers of complexity that goes beyond our perception and comprehension.

The answer here is to prevent by creating the right environmental conditions that keep everything in balance. The quality of the air, water, light and your soil will keep everything in balance.

Building the microflora in your soil is the best way to grow healthy plants that have a healthy immune system to fight off pests and diseases. Microbes produce enzymes and vitamins that break down proteins and feed your plants the necessary elements that your plants need to reach its full genetic potential.

When handling your plants always use sterilized tools and gloves to prevent infection. Do not cross contaminate from garden to garden.

Place filters on air vents coming in and out of your grow room to prevent pests from crawling in.

Foiler spray your plants with compost teas and fermented teas to build up your micro-biome.

Using a fugal dominant tea will help fight those unwanted pathogens. Cut back on the molasses and replace with fish hydrolysate and a pinch of alfalfa meal. Brew for 36 hours. Do not use fish hydrolysate that has been chemically deodorized.

Nutrient Deficiencies

Nutrient deficiencies are identified by a yellowing of the leaves. It is normal for some leaves at the bottom of your plant to turn yellow and it is normal for the leaves to turn yellow during flowering. The less food that is available for your plant the lighter the green will be on your leaves, they will start to turn yellow, develop brown spots and leaf burn. The thing that is different from a nutrient toxicity is that a yellowing of the leaves surrounds the damage.

Some growers, like myself, will allow the plant to feed off itself during late flowering. During this time it is normal for the plant to show signs of nutrient deficiencies. The plant is in its final stages of life and is preparing to die. The plant will shed its leaves, providing food for the next season.

In the event of an unwanted nutrient deficiency, flush the plant out and start on a mild all-purpose nutrient solution that has all the micro and macro nutrients that the plant needs. Once the plant starts to grow in about a week, slowly increase the feeding schedule.

Nutrient Toxicity

The more nitrogen the plant has available to it, the darker the color of the leafs. You don't want your plants to be too dark. You are looking for a light green color. A darker green color is the first sign of toxicity, after that, the leaves will curl under and after that, the tips of the leaves will burn. It can happen easily, especially with salt-based nutrients. Flush the pot out immediately with fresh clean water. Your plants will be in shock and they could take a week before they start growing again, but until they do, just start them on a mild all-purpose solution at a quarter of the recommended strength for growing.

How much nutrients you feed your plants depends firstly on the genetics. Some strains are just more efficient feeders and require less food. Each strain has its own sweet spot for pH, which aids in nutrient uptake. The quality of the air, the available light, the humidity, and temperature all play a roll in determining how efficient your plants can process food. Marijuana can grow up to six inches a day under ideal conditions. Let your plants tell you what they need, start out with a mild solution and slowly increase the recipe until you have the formula for success.

Mold Issues

Mold travels in the air, even the cleanest of grow rooms can get a mold infection. Mold grows in cool wet conditions. White powdery mildew is the most common problem you will be faced with and that problem can be fixed or prevented with a sulphur burner. A sulphur burner will keep your plants alkaline. Mold can't grow in an alkaline environment.

A sulphur burner is basically a tin can and a ceramic light bulb to heat up the sulphur. You will need to operate your sulphur burner on a timer. Run your timer at night, halfway thru the sleep cycle. This is when the stomata of the plant close and the sulphur will not affect the taste of your herb. I run my burner for about four hours for three or four nights to take care of most mold issues.

I prefer not to spray but if I do, baking soda and water are effective at managing mold issues. There are many other types of mold issues like stem rot or bud rot but they can be all avoided by keeping humidity levels down and refraining from over watering your plants.

Bud rot can be prevented by not growing huge colas. If you get bud rot, the only way to treat it is to remove it.

Making sure the nighttime temperature doesn't drop more than ten degrees will prevent mold.

Mold will not be able to establish itself on a healthy plant. Growing organically by building a healthy micro biome in your soil or grow mix will cultivate the microorganisms that will break down the proteins that the mold is made from.

Foiler spray with compost teas to build up those microorganisms that will produce the enzymes that break down the proteins that the mold is made from.

Pest Control

Spider mites are most likely the pests that will infest your garden. Every grower knows how damaging spider mites can be if left unchecked. I use azamax, neem oil and castile soap to control them. I started out with a low dosage of azamax and slowly increased the dosage because those mites will build up a resistance to it. I use about 10ml per gallon of water. As time progressed, I added 5-10ml of neem oil to the recipe and about 15ml of Castile soap.

Canola oil, peppermint, lavender, anything that is an antiseptic will work as a bug repellent.

Those little buggers go thru hell. It does not kill them right away but they stop eating and reproducing and eventually die. The treatment doesn't kill the eggs and they take about three days to hatch, so you are going to have to spray again in about three to eight days.

It helps to have a 10x magnifying glass or a microscope to keep a keen eye on the colony. Once a spider mite hatches it will take about eight days before the females are laying eggs, so time the assault when the eggs have finished hatching and before they start laying down the next generation. You can wipe them out and appear to be pest free for a while but some eggs will lay dormant and they will be back and by the time you notice it, it will be too late, so stay diligent and spray the plants once a week, if you think they need it or not.

I will not spray when the plants are flowering, spraying directly on the buds will affect the taste of your herb. Spider mites eat the leaves, not the buds so what little damage they

do, it will not have a significant impact on your yield or quality if there is an outbreak late in the season.
I sometimes use a vacuum cleaner to suck the mites off the buds but be careful not to damage any trichomes.

The stress of mites may enhance the ripening process since plants are equipped to deal with the day-to-day stresses of its environment.

You can buy predatory spider mites and they will eat your veggie loving mites along with their eggs. Nature will balance out the population of meat eaters and by the way, did you know spider mites are edible. I would not try smoking them but they are a rich source of protein.

If you place a bamboo Cain that is taller then the plant in the middle of the plant or where your top buds are, the spider mites will crawl up the Cain thinking they are climbing to the top of the plant. Whether they are acting under some kind of hierarchy or they are just reaching towards the light, they all get caught up and form some kind of nest on top of the Cain. From there they are easy to suck up into a vacuum or you can eat them if you're hungry. For some reason, they will not crawl back down the Cain onto the plant but they may start crawling out of your mouth if you don't chew them properly.

You can also use azamax and neem oil like a nutrient and feed it to your plants with water. Pests will not enjoy eating your plants and stay away. There are other oils that repel pests and many growers have fun coming up with recipes that work for them. It is good to experiment and keep those bugs on their toes or off their toes. They will adapt, so keep upping or changing the formula if you are not getting the desired results.

Use castile soap to emulsify whatever oils you are using.

You might want to check out my book on herbs and spices. There are many herbal extracts that can be made and used to repel insects.

There is a host of other insects that can infest your garden and more drastic measures may need to be taken. Bug B Gon, (Safers soap) can be effective for a host of unwelcome visitors. It would be wise to alternate methods so mites and other pests don't become resistant to the treatments. There are chemical insecticides available and some claim to break down within 24 hours, but they are toxic and it is wise to avoid using these products. Apart from contaminating your environment, chemicals can easily burn your plants and or send them into shock.

If you have an agricultural license you will have access to far more effective chemicals but they are very harmful to the environment and the finished product. If the plant is healthy it will be able to defend itself. If you have to resort to toxic chemicals then you need to keep reading.

Fungus Gnats are also a common problem, have you noticed any tiny flies flickering around lately? They prefer to crawl around and only fly when disturbed. They look harmless but they can be devastating. Treat these pesky flies with Azamax, Neem Oil, Castile Soap or Bug Be Gone. That is the first step. Next, you have to treat the soil medium because they lay their eggs in the soil and when they hatch, the larvae eat the roots. They can essentially kill your plants in no time. After killing the flies you can treat the soil with Tanlin. For the initial treatment, use two drops per gallon of water and drench your plants. This will kill the larvae in the soil medium, you probably didn't get all the flies so you should do this once a week, using one drop per gallon as a regular maintenance program. You could also use a top dressing to keep those flies down or a repellent that kills mosquito larvae will work.

Use sticky pads and tape to trap the flies. If you use sticky tape around the bottom of your plants this will prevent crawlers from attacking your plants.

Feeding Schedule

When using bottled nutrients I prefer to work on a three-day schedule. I water on the first day, feed on the second day and let the plants dry out on the third day. I find this schedule to work best. Adding a nutrient-rich solution to a wet medium is more forgiving then adding to a dry medium. There will be less chance of shock or plant burn. Some commercial growers will feed for two days on a four-day cycle. To each their own.

Your plants will adapt to you but you have to be sensitive to the plant's needs. If one of your plants is still wet when everything else needs watering, you know the plant has a problem. It is not drinking and feeding and it is obviously sick. Continuing to water that plant my drown it so watch, observe, document and react.

Plant Behavior

This is an area of science where most mainstream scientists fear to tread, in fear of being chastised by their community but there are a few brave souls who will peruse science and not the almighty buck.

Mainstream science does not believe that plants behave because they believe plants do not have a brain. Scientists who are studying plant behavior have not found it yet. They claim that they don't know where to begin to look for a brain but the plant's brain is its root system. The root system is a network, just like the brain that receives and transmits electrochemical information to the rest of its body and its environment. The roots are the plant's brain, the stem is its spine, the branches are its legs and arms. I am not sure if plants have assholes, I would not know where to begin to look.

Plants are motivated by three basic instincts, just like us. They need food and water, they need to nurture and protect there young and they need to procreate. Plants need to grow. Above ground, they will compete for sunlight. They are territorial but below ground their relationships are altruistic. They will help each other by sharing elements gathered from the environment. They will connect with other species like fungi, which will make minerals bioavailable for the plant to feed, while at the same time the plant feeds the fungi carbon. Fungi are connected to all root systems underground. In essence, the plant kingdom behaves as one organism. What we see above ground is a third of what is going on below ground.

Plants have a sophisticated defense system, they consume elements to fortify their immune system and they attract other organisms to protect them and help them procreate. Plants can be very manipulative. They can bear fruit and flowers that people are willing to kill and die for.

They can provide us with food, medicine, and the raw materials to fuel our industrialized economy and all they ask for in return is to grow them. If you think we are manipulating plants, you've got it wrong. Plants are manipulating us. They will establish a new territory and motivate us to protect it. Plants feed us because we feed them.

Plants will put on the Ritz to attract pollinators and they will also attract predators who will eat whatever is eating them. They adapt to their environment.

Plants can perceive their environment more efficiently then we can, especially if they are healthy. Whole plant consciousness.

Plants are autotrophic, meaning plants can feed themselves if they have all the mineral elements it needs. Microorganisms in the soil will make minerals more bioavailable for plant uptake and can increase plant uptake by a thousand times.

Plants do behave and plants are motivated. Trees will cast their seed far enough from their shadow so their young can grow in the light, yet be close enough to feed them thru their root system.

Five Stages Of Plant Development

The Germination Stage

This is the initial stage of growth and occurs when your seeds embryo cracks open and the seedling produces a tap root. This root will fix itself into the soil and push the newborn seedling up and out of the soil surface.

My preferred method of germination is to place the seeds in a cup of water at or slightly above room temperature. You can use distilled water or bottled spring water but I prefer tap water because it contains chlorine, which keeps the seeds sterile.

Some seeds will drop to the bottom right away but most seeds will drop to the bottom within three or four hours. If your seeds have not dropped to the bottom of the cup by then, just give them a little nudge.

Soak your seeds for about fifteen hours. By this time you should see a taproot emerging from the seed. If not, don't worry, pour your seeds on a sheet of paper towel on a plate and cover with another piece of paper towel. Make sure the paper towel stays wet but not soaking wet. Seedlings need warmth, water, and air to germinate.

Your seedlings should germinate anywhere between three and seven days depending on the strain you are growing. Cover your seeds only with a paper towel that has no dyes or chemicals and germinate your seeds in a warm dark place.

Once your seeds have germinated, the taproot should be about a half to three-quarters of an inch long before planting into your grow medium. Be very careful not to break the tap root, plant your seedling about a quarter of an inch into your grow medium. Do not pat the grow medium down, water in the seedling and keep your grow medium wet but not soaking wet or your seedlings will drop off from the mold.

Two embryonic leaves will open outwards to receive sunlight, pushing the empty seed shell away from the seedling. It takes anywhere between 12 hours to 7 days for seeds to germinate. Once the plant has reached this stage it goes into the seedling stage.

The Seedling Stage

After the first pair of embryonic leaves is receiving light, the plant will begin to produce another small set of new leaves. These leaves are different from the last and will have some marijuana characteristics. As the seedling grows, more of these leaves are formed and push upwards along with a stem. The seedling stage can last between 1 and 3 weeks. At the end of the seedling stage, your plant will have maybe 4 -8 new leaves. Some of the old bottom leaves may drop off. Your first leaves will have one blade and as it matures the leaves will develop three, five or even more fan blades per leaf.

The Vegetation Stage

The plant now begins to grow at the rate which its leaves can produce energy. At this stage, the plant needs all the light and food it can handle.

Initially, the plant will develop a strong root system and It will continue to grow upwards producing new leaves as it grows. It will also produce a thicker stem with thicker branches and with more blades on the leaves.

The leaves that grow from the main stalk will be the largest leaves and they are called primary leaves. They contain some cannabinoids but not as much as the secondary leaves, which grow on the branches that grow from the stem.
The plan at this stage of development is to grow roots, the plant will get established in its pot and begin its vegetation cycle. The idea is to gradually increase the pot size as the plant gets bigger, being careful not to allow the plant to get root bound.

The plants will eventually start to show their sex. When they do it is time for the plant's pre-flowering stage. It can take anywhere between 1 and 5 months from seed for the plant to hit this next stage.

The Pre-Flowering Stage

At this stage, the plant slows down in developing its height and starts to produce more branches and nodes. The plant fills out in the pre-flowering stage. During this phase of the plant cycle, your plant will start to show a calyx, which appears where the branches meet the stem (nodes).

It is very important to achieve this stage of development because the internode growth grows closer together forming the structure for big buds. Also during this stage of development, the plant will naturally bush out creating more bud sites.

Nurturing your plants thru this stage of development will greatly increase your yield.

During this stage, it is easy to sex your plants. Females will have white pistils growing out of the calyx and male plants will grow a pod full of pollen.

This theory is unproven but if you start reducing the growing time gradually from eighteen hours a day, it may trigger the pre-flowering stage. Pre-flowering can last anywhere between one and two weeks.

The Flowering Stage

During this stage, the plant continues to fill out. The plant will show its sex clearly. The male plant produces little pods that are clustered together like grapes.

The female plant produces white/cream pistils that look like hairs coming out of a pod. Each of the plants will continue to fill out more and their flowers will continue to grow. It can take anywhere between 8 to 12 weeks for the plant to fully develop its flowers. During this time the male pollen sacks would have burst, spreading pollen to the female flowers.

The male plants will start to pollinate your females in four weeks but if you remove the male plants and the females do not get pollinated, they will continue to grow for a total of eight to twelve weeks depending on the strain you are growing. Flowers without seed is called sensimilla and that is what you want to grow for medicinal or recreational use.

Ripening / Seed Set

The female plant will produce seeds at this point if she has received viable pollen from a male plant. The seeds grow within the female bud and can take anywhere between 4 weeks to 8 weeks to grow to full maturity. The female pistils will change color before the seedpods burst, sending them to the soil below.

Breeders like to collect their seeds before the seedpods burst. Seeds are mature when they are dark in color. Thru selective breeding, you can collect seeds from plants that have the most desirable traits.

If you want to grow sensimilla, you must remove the males as soon as you identify them. If your females do not get pollinated they will continue to grow for an eight to twelve-week cycle.

When the trichomes turn a milky white from crystal clear they are ready to harvest. When the trichomes turn amber, cannabinoids have begun to break down.

Growing From Seed

Back in the day, dealers had buckets of seed lying around the house. If you needed seed, you just had to ask and they would probably give you a hand full of seeds that would fill a baggie, for free.

These days it is a completely different story. People are protective with there strains and seeds are expensive to buy. It is big business and when you are paying 5-10 dollars a seed, extra care must be taken to ensure success.

I have talked about choosing seed from a medical perspective but now we are going to look at it from a growers perspective. If you choose a Sativa strain, the flowering cycle could be as long as twelve weeks as opposed to Indicas who average about eight weeks. Sativas grow tall and lanky with thin leaf blades. Sativas are not popular amongst growers and pure Sativas are very rare. Sativas are best grown outdoors. However, Sativa strains are in high demand at the compassion clubs and dispensaries.

Indicas, on the other hand, are popular and abundant with many strains to choose from. Most strains available to us are hybrids with Indica and Sativa genetics. Strains with strong Indica genetics grow shorter and bushy and are ideal for indoor cultivation. The buds are dense and they can be very high yielding. The strong heavy pain killing body buzz makes it a favourite for growers and users.

I would avoid using feminized seeds or auto-flowering seeds, the genetics are unstable. Use regular seed stock and choose a hearty strain that can adapt to your environment. You can give the same strain to three different growers and they will all turn out different. It is all about the grower, the environment you create and the gardening practices that you use. The plant will adapt to you and your environment and the plant's genealogy will change regardless of who there mom and pop where.

Where you get your seeds is your business but when you do get them, work in a sterilized environment. Store them in a dry, dark, cool place.

If you start from seed, you will not need to feed them, while at the germination stage of development. There is enough food in the embryo to get the plant started. Once the plant reaches its seedling stage, start with a very mild solution (one-eighth of the recommended strength or less). I start them with a flowering solution. The higher phosphorus level promotes root growth.

As the seedling enters the early vegetation stage, I will reduce the bud formula and incorporate a grow solution.

Growing From Clones

When you do start cloning, it is important that your mothers are healthy and pest free to ensure success.

You will get better results if you reduce your nutrient solution and put the mothers on a moderate feed. This will trigger the plant into feeding of the sugars that are stored in the leaves. You want to induce this condition because when you cut that clone it will not have a root system to feed off.

When you plant your clones into your grow medium, dip them into a rooting hormone, shake off the access and plant your clone with one or two nodes underground. You can let your clones soak for 24 hours in water with a small amount of Aloe Vera. You can also dip your clones into the Aloe Vera plant and use it as a rooting hormone.

Saturate your grow cubes or pellets with water. Do not feed your clones in the first week. They do not have roots so it is pointless and if anything, feeding will inhibit growth.

In the second week feed your clones a very mild solution high in phosphorus for root development. I also use Ryhzotonic, which works well as a root stimulator. The solution should be no more than 5ml per gallon of water. You could also use a compost tea.

While in vegetation, your plants require more nitrogen for lush green growth. If your plants are a dark green, you are running a little heavy on the nitrogen. If the leaves curl under and the tips start to burn, you are definitely overfeeding. Less is more.

The number of blades on the leaves will increase as the plant matures into vegetation. Eventually, your plants will grow into the pre-flowering stage. A calyx will appear at each internode and plant nodes will start growing closer together, getting tighter for solid bud formation.

The plant will also fill out during this stage and it usually lasts for one or two weeks. During this stage, you want to be reducing the grow formula and gradually incorporate your bud formula. Make gradual changes to avoid shocking your plants.

When your plants are ready to trip (flower), you reduce the light from 18 hours down to 12 hours a day. Keep feeding your bud formula up until mid-bloom. After mid-bloom, you want to start reducing the formula, forcing the plant to feed off itself, as it does in nature towards the end of the season, when the soil becomes naturally depleted. It is always best to mimic nature. Also during this time, even though you are reducing your levels, you need to boost your potassium levels because the production of phosphorus burns up potassium and potassium is needed for reproduction. It helps to reduce phosphorus and increase potassium during mid-bloom.

After mid-bloom, the plant is preparing to die and this is the time to clean out carbon build up in your plant. If you don't, your weed will burn black. Starving the plant after mid-bloom will force the plant to eat the sugars that are stored in the leaves. The leaves will turn yellow and drop off as your buds begin to set and ripen. It is not normal for your garden to look green and pristine right to the very end. This means your plants are jacked up with chemicals and the finished product will suffer.

High yields start from day one.

In the last two weeks of flowering your plants will shut down sugars flowing to the root system. Microorganisms in the soil will starve and photosynthesis will stop. To keep the plant photosynthetically active, add small amounts of molasses to your mix. This will keep your crop greener, longer.

When you flush your plants in the final week or two, you are removing whatever built up salts that are in your grow medium. How much water you use depends on the size of the pot and how toxic your soil is. In a five-gallon pot, I can easily flush 3-5 gallons of water until my ppm levels are at an acceptable level.

Before I harvest my plants I will take a bud and dry it out in the microwave, using a defrost setting. I will smoke the bud and what I am concerned about is how clean the herb burns. If she burns clean, she is ready for harvest in that regards

When using organic nutrients rather then salt-based nutrients, there is very little salt build up and less of a need to flush your plants. Also, if you are on the right feeding schedule, it will also reduce the need for extensive flushing, since there will be less carbon built up in the plant material.

If you use bio stimulants you can reduce the plants need for nutrients by as much as 50%.

When feeding your plants, less is more. Start out with a mild solution because what the manufacturer recommends is based on optimum growing conditions. Your environment will determine how much food your plants will require, plus some strains are just more efficient feeders and all strains have their own optimum growing conditions to get that bumper crop. All in all, you have to let the plants tell you what they need. Learning how to read your plants is critical.

When growing with salt-based nutrients, you have to be very careful not to overfeed your plants. A big advantage when using salt-based nutrients is that you can measure thru electro conductivity how much nutrients is in your solution.

You can measure and control how much nutrients are going in and what is more important, how much is coming out. I grow in a non-recovery hydroponic system and the runoff goes to waste. If you feed your plants too much, the nutrients left in your grow medium will crystalize and become nonsoluble. The plants will become locked-out, creating a nutrient deficiency and when you add more food, you will create an even more toxic condition.

The runoff should not exceed more than 200ppm more then what went in. I started using salt-based nutrients for the purpose of writing this guide. I started out with the factory recommended feeding schedule and after two feedings I checked the runoff after watering the plants. The runoff was coming out at 1800ppm in some plants, a definite problem, considering my water measures at 200 ppm.

Immediately I flushed my plants and lowered the runoff down to 400ppm. Again the intensity of light, the color of light, the quality of the air, the humidity, CO_2 levels, ventilation, soil condition, feeding schedules, any drastic changes in the environment and your presence all play a roll in how productive your plants will be and how much they can be fed before a toxic condition is created in your grow medium. Find the sweet spot for the environment you created.

Scientists spend many years and dollars coming up with the perfect recipe for your plants, so these one part formulas work well, however, I would use these formulas at a quarter of the recommended dosage, if you are feeding them often in some kind of hydroponic system.

Consider, when growing organically you are feeding the soil, the soil feeds the roots and the roots feed your plants but when using salt-based nutrients you are killing the soil and feeding your plants directly. Less is more. Just feed them what they need for the day. Check runoff for salt build up.

I used General Hydroponics, the industry standard. I used the Flora Nova Grow with good results. It is a one-part formula and plants will do well without adding anything else but you can enhance productivity by using humic and fulvic acid during vegetation and during bud production to treat the medium and improve nutrient uptake.

If you do not overfeed your plants and give them just what they need, so nothing is left behind to crystallize, you will not need to flush until the final two weeks of production. Constantly check your run-off and make sure it doesn't exceed 200ppm more then what went in.

To be able to realize the plants full genetic potential and to achieve a high Brix count you have to start feeding the microorganisms in the soil. Microorganisms will produce biostimulants that feed the roots, which feed the plants and they can do it 1000 times more efficiently then we can, by force-feeding plants nutrients. If you feed the rhizome layer your plants will be healthy, productive and resistant to pest and disease.

Growing For Recreation

The training methods that I am going to share with you will increase yields and provide a consistent product for the medicinal market. However, when you do this, you loose the top bud, the cola, which is the most potent bud on the plant.

Before I started growing medicinal marijuana I would never cut off the top bud. With the top cola intact and as you work down the plant, the buds become less potent, so it will be necessary to grade your crop.

There will be a big difference in potency between the bottom buds and the top buds. Your top buds are AAA. With all this in mind, if you are growing for the recreational market I would not top my plants, instead I would grow a sea of colas in hydroponic beds or if I grew in pots I would super crop instead of top my plants.

Licensed growers are limited to the amount of plants they can grow, so it becomes unfeasible to grow a sea of green. Commercial growers will always go for yield over quality, so if you want the best, grow your own.

When growing recreational marijuana or should I say cannabis, you want to grow the most potent strains you can get your hands on. Strains that is high in THC and strains that are also high in CBN. Cannabinol (CBN) is a degradation product of THC. CBN potentiates the effects of THC, which is why cannabis that has been well cured becomes more potent.

Growing Techniques

At present, I have a grow room where I do my propagation and cloning and from that room, I feed a veg room, which in turn feeds two other rooms that are constantly in the flowering cycle. I clone every two weeks to ensure a steady supply of medicine. I don't like waiting three to four months for my crop to come in. It is a bit of a grind when you are cloning every two weeks you are also cropping every two weeks.

I grow organically and when I do not use salt-based nutrients I cannot measure accurately the electrical conductivity of the nutrient solution and come up with an accurate ppm reading. Growing organically is as much an art as it is a science.

There are many growing techniques that you can use in an indoor grow to increase your yield. Stressing out your plants will increase the production of growth hormones that the plant needs to heal. If you do not overdo it, you can stimulate growth with the aid of different techniques.

Most of the techniques in this book were created to produce a consistent product for the medicinal market. We are looking for a consistent size, which also helps big time in the drying and curing process. We are also looking for some consistency as far as potency goes and you can achieve that to some degree thru training your plants. The idea is to develop a canopy with many bud sites rather than let the plant grow naturally like a Christmas tree.

The top bud has growth inhibitors that prevent the lower branches from growing. The taller the plant gets, the more the lower branches can grow. This is how the plant forms its Christmas tree-like shape, so when you cut that top bud off, all the branches will compete with each other to be the top bud. You end up with a bushy plant with lots of bud sites, especially if you top the branches as well.

When you cut the top bud, the taproot will stop growing, so it is advisable not to top your plants too soon. Wait until the plant is in its mature vegetative state before topping. Some growers will top their plants when they are around six inches tall but it is advisable to wait until they are around sixteen inches tall.

Super Cropping

Super cropping is a method of bending the branches and the main stem. The advantage is you keep the top cola. Choose where you want to bend the stem and squeeze it a little while twisting it back and forth a few times. This will weaken the inside of the stem allowing it to break and bend over. Do not break the outside layer where the veins of the plant are.

Your plant will recover quickly, growing side shoots, while bushing out. This can be practiced throughout the vegetative cycle and up in till two weeks before flowering.

Topping

Topping simply means cutting the top of the plant. The top bud has growth inhibitors that prevent the lower branches from bushing out and taking over. When you cut the top, your side branches will grow out and up. You can also top your side branches to create more bud sites in the top canopy. Topping will increase your yield and it will also give you a more consistent product for the medical market.

Some growers will top their plants when they are about six inches tall. They will remove the bottom two nodes and allow the plant to grow. When the branches are long enough they will top or super crop the branches. Topping and super cropping work very well together. I get the best results when I use the two methods together.

You can get really strict with your topping by removing the first two internodes and growing the next two internodes. Top again after two internodes and repeat the process until you have the desired effect. Stop topping at least two weeks before flowering.

Lolly Popping

Lolly popping is about removing the lower branches. Doing this will provide more food energy to the top of the plant and yield larger buds. Remove all internodes that are growing below the canopy. There will be less popcorn bud to clean up during harvest time.

Remove everything that is not in the optimum growing zone. Remove those little internodes that develop on the branches below. This will also keep the plant healthier and away from the soil. The plant will also be easier to water.

Monster Cropping

Monster cropping is about cloning a plant when it is two weeks into the flowering cycle. It will take a few weeks to root and revert back to vegetation mode but when it does, many shoots will grow into a bush and you will probably have to thin it out a little. It takes awhile to get started but when it does, growth is very prolific.

Scrogging

Scrogging is all about training your plants to grow horizontally with the aid of a net or some kind of trellis system. As the plant grows, weave the branches in and out of the trellis. This will form side branches, which increases bud sites. This method works very well and with a little patients, you can greatly increase your yield.

With the new height restrictions, scrogging is the way to go. Top your plants when they are about sixteen inches tall and grow them out. Put them in big shallow pots and they could yield a pound of finished product per plant.

Thinning And Pruning

Even though you want to grow a bush with many branches and bud sites you also need to provide room for light and air to penetrate the canopy, so thinning some branches and removing some primary leaves may be necessary for air circulation and light distribution.

When your leaves turn yellow that means the plant is eating the sugars stored in the leaves and that is a good thing, so leave them on until they are ready to drop.

Do not prune more than thirty percent of the plant at any given time. However cutting back your plant will in vigour new growth.

When you start cloning, you are going to want to keep good healthy mothers around rather than clone from clones because you do experience some genetic degradation.

Mothers can quickly get out of hand so it will be necessary to cut them back once in awhile. Also, you will probably have to cut back the roots as well and transplant into a larger pot.

Topping Your Roots

If your roots become root bound take the plant out of the pot and cut back the roots. No more than thirty percent. This will invigorate new growth inside the root ball where your plant can feed more efficiently. A less invasive way is to use fabric pots or drill small holes into the pot so the roots will die off when they come in contact with the air. This will invigorate new root growth inside the root ball.

Fimming

Fimming is just like topping, only you make your cut halfway thru the top bud. Growth will bush out and develop many bud sites. Some thinning may be needed.

Agro Sonics

This may be cutting edge or this may be over the edge but I have recently started treating my plants with sound. A couple of scientists in Korea have done some experiments with it. We have all heard about people talking to their plants and it is considered a little eccentric. Most of us believe that music can enhance plant growth. The truth is, plants respond to sound, they are vibrational beings, much like us.

My son said that if I play rap music half way thru the bud cycle it will increase my yield by 10% but I think he was just pulling my leg.

These scientists in Korea have discovered that certain frequencies stimulate plant growth by as much as 100%. Those frequencies are 125hz and 250hz. There are free sound tone generators online that you can use to get started with if you want to experiment with it. I just started treating my plants with sound, if I see a significant difference in growth, I will probably invest in a sound tone generator or just use one of my synths. I will keep you up to date on my progress.

There is more. Sound can be used as a pesticide to deter bugs from hanging out in your garden. The government is developing sound as a weapon and some frequencies can be very dissonant. Sound can also potentially be used to stimulate nutrient uptake and be used as a source of food.

Sound is energy at lower frequencies than light and just like light, sound can feed your plants. Perhaps the frequencies of sound can be used to target and stimulate the growth of cannabinoids in your plants.

Who knows what the future will bring?

Yield

Assuming that you are on the right feeding schedule, your yield will depend mostly on genetics and the environment. Your strains will adapt to your environment and develop phenotypes that grow well in your garden. These are the plants that you want to propagate or clone. Cloning from clones is possible and I am sure it is done all the time. It takes many generations before the plant's genetics begin to break down.

Focus on growing soil by feeding the microorganisms in your medium.

Thru vigorous plant training, you can greatly increase your yield in an indoor grow op. Training your plants in wide pots develops a wide canopy with many bud sites. You can increase your yield from one ounce a plant to a pound a plant or more with the right growing techniques.

Size matters, the larger the plant, the larger the yield. The bigger the root, the bigger the fruit. When growing indoors, you are limited to a canopy that is 12-24 inches thick so growing tall plants indoors doesn't work well. Keep them short and bushy.

Growing no-till style will not only increase yield but it will also increase quality as well. Usually if you go for yield you sacrifice quality or if you go for quality you sacrifice yield but with the no-till method it is a win win situation. You gain on quality and yield.

The Outdoor Grow

Marijuana is a weed. It will grow just about anywhere. Marijuana does very well outdoors and less care is needed to grow big healthy plants.
Nature has a way of providing. The natural sunlight will produce a richer cannabinoid profile. Regardless of what some growers claim, you can grow good quality herb outdoors.

It is easier and cheaper to grow outdoors, but there are security issues to overcome. There is a good chance of getting jacked or popped.
The growing season is long and you are looking at one crop a year. That is way too long to wait for medicine.

Growing indoors gives you control over the environment and the genetics. You can produce multiple crops a year which provides a steady consistent supply.
Plant requirements are pretty much the same as growing indoors. Let the plant tell you what to do. It is best to grow no-till style in the ground if your soil is well cultivated or in 600 or 1000-gallon fabric pots.
Your growing season will start in early spring when the soil heats up to about 80°F. and go to about mid-October depending on the strain and the weather, mostly.

Outdoor plants can grow very large and they can yield 5-10 pounds so give them lots of room to breathe.

Some thinning and training will be required much like an indoor plant but not to extreme. You will have to build some kind of trellis system to hold the branches up. Other than that let Mother Nature take its course.

We feed the microbes, the microbes feed the plant, the plant feeds us. The nutrient cycle is complete. The cycle of life. The plant also feeds microbes when they are needed in the mix so plants can be self-sustaining in the presence of the right elements.

Choosing The Right Strain

Choosing what plants to propagate is a question of what traits you consider desirable. The average grower would want to grow a hardy plant that is easy to grow and grows fast. A plant that grows short and bushy for indoor cultivation and a strain that is disease resistant. A strain that roots fast and produces high yields.

You would also want to choose a strain for its taste and aroma and also choose a strain with the right cannabinoid profile for your needs.

It would be a good idea to grow a few different strains, one can become resistant to the same strain over time.

There are many medicinal reasons why you would pick one strain over the other. For medicinal reasons you will need strains with a one to one ratio, meaning there are equal amounts of THC and CBD. Also look for strains that are high in CBN because that will increase your herbs potency during the curing process.

Choosing a strain with high levels of THC is ideal for recreational purposes or for pain relieve.

Every strain is different with different qualities, so choosing the right strain is subjective.

Propagation

Ordinarily, if you grew from seed, you would have to sex the plants and separate the males, otherwise, they will pollinate your females and they will start to set seeds in about 4 weeks into the bud cycle.

If you want to grow sensi you have to remove the males. The male flowers produce little pods full of pollen and they are easily distinguishable from female flowers.

Some growers want to grow seeds. They like to experiment, cross-breeding strains. You can grow male plants in your vegetation room but as soon as they are flowering they have to go in their own room that is sealed with its own independent ventilation system.

Select the plants you wish to propagate and move them into the propagation room on a 12hr flowering cycle. Time it so your females is about 4wks into their flowering cycle before you introduce them to a male plant that is ready to pollinate. This will allow enough time for the seeds to mature.

The seeds will be a dark spotted color when they are mature. Simply air-dry them and they will lay dormant until they are germinated.

Harvest Time

It is harvest time when your buds are mature. The leaves are turning yellow and the buds have ripened. The trichomes that cover the bud turn from crystal clear to a milky white color. Some growers take it until the trichomes turn amber but the cannabinoids in your herb are starting to deteriorate at that point.

Drying

There are a few ways of going about pulling your crop down. One way is to remove the primary and secondary leaves while the plant is still alive, then cut the entire plant and hang it upside down in a dark well-ventilated room.

Some growers will trim the sugar shake before the bud dries. Others will just remove the fresh bud of the stem and use a machine to trim their bud. I have used a machine for a while but I have stopped using it because you loose allot of crystal (trichomes) in the process. If you trim by hand you will have a much better product.

Other growers will simply cut the entire plant or branches and hang them to dry in a dark well-ventilated room.

The temperature of the room should be between 65°-75°F. If the room is too cool the bud will take too long to dry and mold could set in. If the temperature of your room is too hot your bud will dry out too fast and it will affect the taste in a bad way.

The humidity in the room should be between 45-55%. If the humidity is too low your bud will dry too fast and the taste will be affected. If the humidity is too high, your bud will take too long to dry and your bud will be susceptible to mold.

Keep the air circulating in the room by placing a fan above your hanging plants. The herb should take about two weeks to dry. When you bend a stem and it breaks with a snap, it is dry enough to cure.

Bone the plant by removing the bud off the stem. Manicure the bud by removing the sugar shake from the bud. One method is to remove the bud from the plant and trim the sugar shake while the buds are fresh. Once the buds are dry they are ready for a final manicure.

When trimming your bud by hand or manicuring, work on a glass table so you do not waste the kief that falls off the bud. Once the green material is removed, you can scrape off the kief and compress it into hash. Yum.

Wear latex gloves, they will get covered in resin. When they do, freeze them and just give them a snap to remove the resin. Roll it up into a ball of hash and smoke it or use it for edibles.

Keep your tools clean. Scrape of the resin and harvest the hash.

When your buds are dry, they should have lost about 70% of its fresh weight.

Curing

Curing is an art unto itself. It takes practice to get it right. You can cure your bud in paper bags, plastic bags or glass jars. I prefer glass jars over plastic even thou both have their advantages. For a better quality product, use glass jars. If you are curing large crops, you could rig up an aquarium with an airtight lid of some kind.

Some people claim they can taste the plastic in the finished product and from a scientific perspective, cannabinoids like to attach themselves to the vinyl in the plastic. It is a good idea to not use plastic.

The trick here is to get the right moisture content before they go into the jars. When finished, your buds should have about 10-12% moisture content.

Grade your bud into small, medium and large sizes. Cure your buds with a consistent bud size in each jar, so they cure at the same time. If your bud starts to sweat up and moisture appears on the inside of the jar, or if your buds smell like formaldehyde, then your buds are too wet and you need to remove them from the jar and dry them out a little more.

Fill the jars but do not pack them too tight. You need some air in the jar, but not too much for the fermentation process to happen. Open the jar and air the buds out for a few minutes, several times a day or when needed for the first few days. As the bud cures, it gets drier and you do not need to burp your weed as often.

You can get a nice cure in about two weeks, but the longer you leave it the better it gets. Two months would be nice if you can hang on to it that long. As starches convert to sugars, the herb gets sweeter.

Also in time as the herb cures, the CBN in your herb will enhance the THC content and your weed becomes more potent. Most growers will not get their herb to this level, but it is well worth the effort and time that goes into it.

Some people will cure their herb for years like a fine wine. You won't get that kind of quality at the local compassion club, but if you grow it yourself and cure it properly, you will end up with a product that connoisseurs can appreciate.

Cannabis

Cannabis is the cured dried flower of the marijuana plant, commonly known as bud. Most people will smoke or vaporize the bud for a euphoric effect but most of its medicinal properties deteriorate in the process.

 A more efficient way to medicate is to ingest cannabis by making natural remedies like cannabis oil, hash, tinctures, canna oil, cannabis paste, and cannabis butter. From these extracts and infusions, you can make an endless list of marijuana edibles.

The Medical Marijuana Guide. NATURES PHARMACY. Shows you how to make cannabis infusions and extractions. The guide also has many recipes and also covers the benefits of eating fresh raw marijuana.

People smoke or vaporize cannabis. Both methods of smoking cannabis are just as effective medicinally, with vaporizing there is fewer carcinogens to deal with. You can also control the temperature of the burn and target certain cannabinoids with a vaporizer. Know your numbers.

Regardless of smoking and vaporizing inefficiencies and efficiencies, smoking or vaporizing cannabis has many health benefits. Smoking cannabis is very therapeutic, even when smoked recreationally.

Marijuana does not stop becoming a medication when consumed for recreational reasons.

Smoking or vaporizing cannabis will not correct the imbalance that caused your symptoms to begin with, but smoking cannabis provides instant relief and the dosage is controllable. Smoking or vaporizing provides instant relief for a wide array of symptoms.

There are many hybrid strains of Cannabis to choose from and they are basically Indica, or Sativa crosses. Pure Indicas are sedative and numbing, while pure Sativas are uplifting and creative.

For medicinal purposes, you are looking for a strain that is rich in CBD. a non-psychoactive ingredient with powerful health benefits.

Some strains like Charlottes Web or Cannatonic contain higher levels of CBD, which are highly sought after for serious illnesses. CBD is a regulator and will identify THC as a toxin and balance out its narcotic effect on the body.

CBD Rich Strains

Critical Mass. 5% CBD

CBD Skunk Haze. 5% CBD

Nordle. 5.5% CBD

Shark. 6% CBD

Charlottes web. 17-23% CBD

Cannatonic. 25% CBD

These numbers are estimates. The numbers will change from grower to grower, which depends on the environment and how they were bred and raised. Most or all CBD rich strains are Cannatonic strains developed in Amsterdam.

Indica or Sativa

The effects of smoking Cannabis is euphoric, sedative and can vary from strain to strain. There are many strains out there and for medicinal applications, you need a strain that is rich in CBD. CBD is a regulator and will reduce THC content.

There are two basic strains that are commonly used and there are many hybrids of the two strains. There is a third strain Ruderallis, it is adapted for winter climates but the THC content is very low.

The most common of the two strains is Indica. Indica strains have broad leafs and are short, fat and bushy with a short growth cycle. Indoor growers prefer growing Indica or a strong Indica / Sativa hybrid.

Sativa has thin leaf blades, which has a longer growth cycle and grows up to twenty-five feet in a season. Indoor growers will avoid growing Sativa but there are hybrids that have adapted for indoor growing. A pure Sativa strain is rare.

Each strain has its own range of effects on the mind and body, resulting in a wide range of medicinal benefits.

The effect produced from smoking Indica bud is a strong physical body effect that will make you sleepy and provides a deep relaxed feeling, compared to the Sativa effect, which is known to be more energetic and uplifting. (meaning you sleep lighter)

Because Sativa and Indica strains have very different medicinal benefits and effects, certain strains can be targeted to better treat specific illnesses.

Indica dominant marijuana strains tend to have a strong sweet, or sour aroma to the buds (ex. Kush, OG Kush) providing a very relaxing and strong body high that is helpful in treating general anxiety, body pain, eating and sleeping disorders.

Medical marijuana patients smoke Indica in the late evening, or right before bed due to how sleepy and tired you become when high from an Indica strain like Kush.

Benefits of Cannabis Indica

1. Relieves body pain
2. Relaxes muscles
3. Relieves spasms, reduces seizures
4. Relieves headaches and migraines
5. Relieves anxiety or stress

Sativa dominant marijuana strains tend to have a more grassy type odor to the buds, providing an uplifting, energetic and cerebral effect that is best suited for daytime smoking. A sativa high is one filled with creativity. Many artists take advantage of the creative powers of Cannabis Sativa.

Benefits of Cannabis Sativa

1. Feelings of well being and at-ease
2. Uplifting and cerebral
3. Stimulates and energizes
4. Increases focus and creativity
5. Alleviates depression

Cannabis stimulates your appetite, so if you are obese you can curve those food cravings with a glass of water. If you have many snacks during the day instead of three large meals you could actually eat more and loose weight.

If you are anorexic you won't be for long.

Over all, cannabis has many health benefits and I find it therapeutic even if just smoked for recreational reasons. I find that smoking smaller amounts (less more often) leaves me more energetic and alert.

Cannabis Root

Cannabis root contains very high levels of CBD, which makes it a very valuable medicine. There are many ways that it can be prepared. It can be dried out and ground into flour. From there you can make paste or salves. It can be also consumed raw. It can also be juiced or added to smoothies. You can also make tea out of the dried root.

Cannabis root was used extensively as a medicine by the ancients. It is also used to counteract the effects of THC. If you underestimate an edible or you just lose control and get the munchies, cannabis root can be used to counteract the effects of THC. Cannabis (the dried, cured flower we know as bud can contain very little CBD when tested. Cannabis root can be used to fortify your infusions and extractions. The root could also be used to regulate levels of THC in your medicine.

Add 2 tsp. of cannabis root to a smoothie and wait for about an hour for it to kick in. Your pain will melt away leaving you feeling relaxed without a heavy THC buzz. You will feel perfectly medicated for most of the day, feeling energetic, happy and feeling good for daytime activities.

In much of the world, the main methods of preparation have been consistent through time. The root is either applied raw, dried, boiled, soaked, roasted or occasionally reduced to ash. The ancients believe that cannabis root ash and honey can rejuvenate hair growth when applied to the scalp.

If boiled for a short time it can be drunk as a tea, if boiled for a long time it reduces to a thick, dark extract resembling pitch or heavy oil. If it is dried or roasted and ground, it forms a powder that can be rendered into salves or poultices; soaking it can produce a soothing, moist bandage for inflamed, burned or irritated skin.

Marijuana and Spirituality

When I talk about spirituality I am not referring to peoples religious belief system. I am referring to the energy that flows thru our body that creates pure consciousness. Whatever you are, Christian, Muslim, Hindu or Buddhist this energy flows thru us all, it is universal.

We are vibrational beings of pure consciousness, generating and working with energy to shape our world.

We are vibrational beings and marijuana plants are also vibrational beings that manifest's energy. Energy is basically a language like music is language. Marijuana resonates on a multitude of frequencies that is harmonious to our electromagnetic field. Marijuana opens our chakras. Chakras are like gateways in your body that open and close allowing energy to flow in and out.

After smoking marijuana, our chakras can double in size, providing us with twice as much energy to heal.

Spiritualists refer to the marijuana plant as a knowledge plant, a teacher plant and that is because the plant's energy field contains electrical and chemical information that brings us into a calm, peaceful state of awareness.

Marijuana reduces resistance. Resistance creates tension, tension creates stress, stress creates pain and stress inhibits the flow of vital energy in our body.

Marijuana is a sedative that relaxes you, the tension is released and the pain melts away. A body at ease is free of dis-ease.

When the mind is calm and the body is relaxed the mind and body begin to heal. When the mind and body are relaxed, it is open to the healing energy that abounds us.

There are many chakras located thru out the body, but what we should be primarily concerned about is the seven primary chakras. Each chakra channels energy to a different system in the body.
If a chakra is congested or depleted it will have a negative effect on the system it feeds energy too. Too much energy can be just as crippling as not enough energy and the objective is to always maintain a balance.

A healthy chakra system creates a strong energy field that will protect you from your environment. This energy flows outside of the physical body and is your body of light. We are beings of light and our physical body is a vibrational manifestation of that light. You are the light and you can expand several feet beyond the physical plane of existence.

Modern science can now measure this aura of light that we manifest. We can also take pictures of it. Energy healers know that disease shows up in the aura long before the disease manifests itself in the physical realm.

When your body of light becomes depleted you become vulnerable to disease. Using marijuana can be very effective at restoring balance and restoring the levels of energy in your body.

This energy is well known as Chi or Qi. Different cultures have different words for it. Science identifies it as subtle energy and they are just beginning to understand its true nature. Subtle energy can travel 400 light years in about a quarter of a second. Try wrapping your mind around that.

Starting at the base of the chakra system is the root chakra, which resonates with the color red. The root chakra primarily channels energy into the eliminative system.

The second chakra is the sex chakra, which resonates with the color orange. The sex chakra primarily channels energy into the reproductive system. It is here where life is created.

The third chakra is the solar plexus and resonates into the color yellow and primarily channels energy into the digestive system.
The fourth chakra is the heart chakra and resonates with the color green. The heart chakra primarily channels energy into circulation.

The fifth chakra is the throat chakra and resonates with the color blue. The throat chakra primarily channels energy into respiration.

The sixth chakra is the third eye chakra and resonates with the color indigo. The third eye chakra primarily channels energy into cognition, the endocrine system and the central and peripheral nervous system.

The seventh chakra is the crown chakra and resonates with the color white. White is a reflection of all the colors in the spectrum and channels energy into the immune system. This light can also appear to be violet or purple. White light can also channel energy thru out the entire body and can be used to treat all the systems in the body.

I hope I have helped you gain a basic understanding of how energy works in the body and how marijuana can enhance the mind and bodies ability to make a recovery. I am a Qi Gong practitioner, working with energy is kind of my thing.

Marijuana And Addiction

So far, I have talked about the many benefits and the many ways you can use marijuana as a medicine and a complete food source. Medical marijuana is saving peoples lives. No doubt about it, putting the political, legal and economic side of things aside, marijuana is improving the quality of life for many people.

With that said, I can't help but feel responsible for looking at the big picture. I must be unbiased and see both sides of the coin, so I asked myself, I said self, what are the pitfalls associated with using marijuana? Now I wasn't too highly motivated to go there but my thirst for knowledge is unquenchable.

Can marijuana be addictive? There are different levels of addiction. Marijuana isn't biologically addictive like heroin but you can develop a dependency on marijuana that starts out as pleasure-seeking behavior but ends up as obsessive-compulsive behavior.

Marijuana can increase your levels of dopamine by tenfold, leaving you feeling very euphoric.
Your brain is a communications network and the front part of your brain is called the cerebral cortex, which receives information from your senses, processes them and sends chemical and electrical messages to other parts of the brain. From the cerebral cortex, the information travels towards the brain stem where your Medulla is located.

The Medulla has many functions and one of them is to determine right from wrong based on survival needs. If you do something right you are rewarded with an electrical impulse and a chemical message that releases chemicals like dopamine in your limbic system. The electrical impulse reinforces the behavior by motivating you to repeat the behavior.

When you do not achieve or accomplish your goals or if your behavior has bad consequences your behavior is punished with chemicals like adrenaline to discourage your behavior. It is all about survival and this is how the brain works. It is a punishment and reward system that motivates behavior much like Pavlov's theory of conditioning.

If you use a stimulant like marijuana you are basically hijacking your Limbic system. After prolonged use, you lose the ability to enjoy life's simple pleasures, the motivation drops, depression sets in and you are now using to feel normal.

To avoid developing a marijuana dependency, practice moderation and it is a good idea to dry out once in a while. Stop using for awhile and allow your brain to get back to normal naturally. Smoking recreationally is one thing but when you start smoking marijuana every day, it is going to have some negative consequences that you need to watch out for.
If you are a medical user, then your dependency is justified. You need the medication to make a recovery. Medical marijuana is far less invasive then most pharmaceutical medications and medicinal marijuana can be more effective in most cases. You have to weigh the pros and cons and decide for yourself if medical marijuana will benefit you or diminish you.
When you are dying of cancer or you are in excruciating pain, medical marijuana can benefit you as a narcotic.

There are also many ways to medicate with medicinal marijuana without the psychoactive effects that lead to compulsive behavior.

You can avoid decarboxylating your herb or you can consume medical marijuana in its fresh raw form, which is completely non-psychoactive and you still get the medicinal benefits, other than the euphoria. The feeling of being healthy and vibrant again feels better than a euphoric feeling that is blanketing an underlying condition.

Because marijuana is fat soluble, it takes longer to withdraw but it is easy to manage compared to a tobacco, alcohol or a heroin addiction.

Four percent of the population suffers from addiction. That number has not changed thru-out time. That means that there is nothing culture can do to encourage or discourage addiction. We can leave bowls of cocaine at the end of every driveway and that number will not change.

Medical Marijuana

The use of medical marijuana has been well documented for over eight thousand years. Ancient civilizations have evolved and thrived around the marijuana plant. The marijuana plant has commercial, industrial, medical, nutritional and recreational benefits. All you need is weed.

After several decades of research, scientists studying the effects of marijuana made several important discoveries. Not only did they identify the active ingredients in marijuana, they also discovered where and how they work in the brain, via a system called the **endocannabinoid system.**

The endocannabinoid system is a communications network located in the brain and body that affects important functions, including how a person feels, think, and behave.

The natural chemicals produced by the body that interact with the endocannabinoid system are called cannabinoids and like THC and CBD, they interact with receptors to regulate these bodily functions.

How Cannabinoids and Neurotransmitters work

Brain cells (neurons) communicate with each other and with the body by sending chemical messages. These messages coordinate and regulate everything we feel, think and do. These chemicals (called **neurotransmitters**) are released from a neuron (a presynaptic cell), move across a small gap (the synapse), and attach to receptors located on a nearby neuron (postsynaptic cell). This triggers a set of events that passes the message along.

The endocannabinoid system communicates in a different way. It receives chemical messages. When the postsynaptic neuron is activated, cannabinoids, (chemical messengers of the enndocannabinoid system) are made on demand from lipid precursors (fat cells) already present in the neuron. They are then released from that cell and travel *back and forth* to the presynaptic neuron, where they attach to cannabinoid receptors.

Since cannabinoids act on presynaptic cells, they can control what happens next when these cells are activated. In general, cannabinoids function like a regulator for presynaptic neurons, regulating the amount of neurotransmitter (e.g., dopamine) that is released, which affects how messages are sent, received and processed by the cell.

Medicinal marijuana feeds our endocannabinoid system, whether we like it or not, we are all hard wired to thrive on cannabinoids.

The body naturally produces cannabinoids, but with the onslaught of the industrial revolution, the fast food industry, chemical insecticides and pharmaceutical drugs have overwhelmed our immune system. Our bodies can't keep up with the constant exposure of toxins that don't belong in our body.

Fortunately the marijuana plant produces abundant amounts of cannabinoids that our endocannabinoid system thrives on. The brain will begin to communicate and begin to holistically restore physiological balance. The body begins to heal and will prevent as well as cure dis-ease.
There is substantial research by many scientists who can prove that medical marijuana can alleviate many different diseases.

We are learning from these scientists, but when it comes down to it, marijuana does not treat the disease, marijuana treats the body. The body treats the dis-ease, naturally, as nature intended.

Cannabinoids treat our symptoms by identifying the problems and fixing it.

Isolating marijuana compounds and marketing them as medications is not very effective. Like most pharmaceuticals they have harmful side effects that could result in death. Marijuana compounds interact and respond to your bodies needs. Isolating compounds is like creating a symphony orchestra with one instrument, or trying to ride a bike with just one wheel.

Marijuana in its natural form is a very powerful medicine with no known harmful side effects.

Marijuana, is not just an effective natural medicine, it is also one of the most nutritious plants in the world. Marijuana is a complete protein with ten amino fatty acids. Marijuana is a "super food."

Top Ten Health Benefits

1. Cancer
Cannabinoids inhibit tumor growth and kill cancer cells. Our governments have known this for some time and they continue to suppress the information.

2. Tourette's Syndrome
Tourette's syndrome is a neurological condition characterized by uncontrollable body movements.
Dr. Kirsten Mueller-Vahl of the Hanover Medical College in Germany investigated the effects of cannabinols in 12 adult Tourette's patients. A single dose of the cannabinol produced a reduction in symptoms for several hours.

3. Seizures

Fresh Raw Marijuana is a muscle relaxant with "antispasmodic" qualities that have proven to be an effective treatment for seizures. There are countless cases of people suffering from seizures that have been able to function better through the use of fresh raw marijuana.

4. Migraines
Since the legalization of medicinal marijuana California doctors have been able to treat more than 300,000 cases of migraines without the use of conventional medicine. I do not get migraine headaches when I eat fresh raw marijuana

5. Glaucoma

Medicinal marijuana's treatment of glaucoma has been well documented. There isn't a single valid study that disproves marijuana's powerful effects on glaucoma patients.

6. Multiple Sclerosis

Former talk-show host, Montel Williams began to use medicinal marijuana to treat his MS. Medicinal marijuana inhibits the neurological effects and muscle spasms, symptoms of the fatal disease.

7. ADD and ADHD

Medical marijuana is a perfect alternative for Ritalin and treats the disorder without any negative side effects.

8. IBS and Crohn's

Marijuana has shown that it can help with symptoms as it stops nausea, abdominal pain, and diarrhea.

9. Alzheimer's

The Scripps Institute, in 2006, proved that THC found in marijuana works to prevent Alzheimer's by blocking the deposits in the brain that cause the disease.

10. Premenstrual Syndrome

Medical marijuana is used to treat cramps and discomfort that causes PMS symptoms. Using marijuana for PMS goes back to the day of Queen Victoria.

Treating the disease or the symptoms is a clinical approach. Treating the body where the disease lay is a holistic approach. When the body is in balance the body will regenerate and heal the disease.

Raw Marijuana

Mounting Evidence Suggests Raw Marijuana is Best.

Cannabinoids found in marijuana can prevent cancer and reduce heart attacks by 66% and insulin dependent diabetes by 58%. Cannabis clinician Dr. William Courtney recommends drinking 4 - 8 ounces of raw bud and leaf juice from any hemp or marijuana plant. 5-10 mg of Cannabidiol (CBD) per kg of body weight is recommended by the FDA.

Why raw?
Heat destroys enzymes and nutrients in plants. Fresh raw marijuana allows for a greater availability of these elements. Those who need large amounts of cannabinoids without the psychoactive effect can utilize the medication at 60 times more tolerance than if it were heated.

Raw marijuana is considered to be by experts, a dietary essential. A powerful anti-inflammatory and antioxidant, raw cannabis is a super food, a power food, food that is functional, food that heals, Natures Pharmacy.

You can juice the leaf and or bud. Drink it straight or mix it with other juices. Making cold-pressed juice is recommended. You can add fresh raw marijuana into your smoothies or you can add it to your salads. Please refer to my Medical Marijuana Guide. NATURES PHARMACY for more information.

Good health begins in the gut.

Cannabis Extracts

Cannabis contains over four hundred medicinal compounds, which create a beneficial synergy that becomes something greater than the sum of its parts. Meaning they work together to enhance or inhibit endocannabinoid activity and their effects on the mind and body.

Cannabis is a herb and the essential oil extracted from the plant can be three to four times more potent when metabolized thru the gastric system.

Be careful when making or buying extracts. The finished product should be a thick tar-like substance. If the oil is too thin it contains water or alcohol.

Delta 9-Terahydrocannabinoil (THC)

9-tetrahydrocannabinoil (THC) ingested orally undergoes "first pass metabolism" in the small intestine and liver forming 11-hydroxy THC; the metabolite is more psychoactive than THC. Inhaled THC undergoes little first-pass metabolism, so less 11-hydroxy THC is formed. Smoking cannabis is an expedient in fighting fatigue, headache, and exhaustion whereas oral ingestion of cannabis results mostly as a narcotic effect. THC-9 boils at 157°C and its properties are Euphoriant, Analgesic, Anti-inflammatory, Antioxidant and Antiemetic.

Cannabinoil (CBD)

Cannabidiol (CBD) is the next-best studied phytocannabinoid after THC. CBD works in synergy with other compounds as a regulator.
CBD provides antipsychotic benefits. It increases dopamine activity, serves as a serotonin uptake inhibitor and enhances norepinephrine activity. CBD protects neurons from

glutamate toxicity and serves as an antioxidant. THC inhibits receptor activity in the hippocampus and effects short-term memory. There is a good reason for this but CBD, on the other hand, does not dampen the firing of hippocampal cells and does not disrupt learning. CBD works in synergy with other cannabinoids and is more effective medicinally then THC alone. CBD is non-psychoactive and has little effect on the immune system. CBD boils between 160-180°C. and its properties include Anxiolytic, Analgesic, Antipsychotic, Anti-inflammatory, Antioxidant and Antispasmodic.

Cannabinoil (CBN)

Cannabinol (CBN) is the degradation product of THC. CBN potentiates the effects of THC, which is why marijuana that has been well cured is more potent. CBN boils at 185°C and its properties are oxidation, breakdown, product, sedative, and antibiotic.

Cannabigerol (CBG)

Cannabigerol (CBG) works in synergy with other cannabinoids and does more or less what other cannabinoids do but CBG has far superior antibacterial and antifungal properties. CBG boils at 220°C and is Anti-inflammatory, Antibiotic, and Antifungal.

Terpenoids

The unique smell of cannabis arises from over one hundred terpenoid compounds. Terpenoids act on receptors and neurotransmitters as a serotonin uptake inhibitor. Terpenoids enhance norepinephrine activity, increase dopamine activity and augment GABA receptors.

Inhaling these synergistic terpenoids reduces anxiety and depression, which improves immune function via the neuroendocrine system by damping the hypothalamic-pituitary-adrenal axis.

Inhalation of terpenoids reduces the secretion of stress hormones and is very sedative. Inhaling terpenoids increases cerebral blood flow and enhances cortical activity. Terpenoids are very effective at treating Alzheimer's disease. Terpenoids boil between 119-224°C. Their properties are Analgesic, Anti-inflammatory, Cytoprotective, Antimalarial, Cannabinoid agonist, Immune potentiator, Antidepressant, Antimutagenic, Sedative, Anxiolytic, Immune potentiator, Antipyretic, Increases cerebral blood flow, Stimulant, Antiviral, Antinociceptive and Bronchodilator.

Flavonoids

Flavonoids are aromatic, polycyclic phenols. Cannabis consists of about one percent flavonoids. Flavonoids have a wide range of biological effects including many properties found in terpenoids and cannabinoids. Flavonoids have a high affinity for estrogen receptors. Flavonoids boil between 134°-250°C. Flavonoids properties include Anxiolytic, Anti-inflammatory, Estrogenic, Anti-oxidant, Antimutagenic, Antiviral and Antineoplastic.

Refer to my book The Medical Marijuana Guide. NATURES PARMACY to learn how to make extracts and infusionsco

Know your numbers

The idea is to make your infusions and extractions as potent as possible. This way you can use as little as possible in your recipe with not much effect on the taste of your finished product.

Be aware when ingesting cannabis, THC9 converts into THC11, which is more potent and longer lasting.

When calculating how much extract or infusion to use in your recipe, start out with how potent your herb is. If your herb had a THC content of 14% and you want to fuse 28 grams of marijuana into 454 grams of butter, convert the dry weight of your marijuana into milligrams. One gram of dry herb is equal to one thousand milligrams so 28 grams is equal to 28,000 milligrams times 14% is equal to 3920 milligrams of THC in 454 grams of butter. 3920 divided by 454 equals 8.634 milligrams of THC in a gram of cannabutter. The FDA recommends 5-10 milligrams per 100 kilograms of body weight. For an average person, I would say 3-7 milligrams is recommended but for first timers, I would cut that in half and see how that goes. It can take up to two hours for your body to metabolize THC, so be patient, patient. Everyone's sensitivity is different and it is best to start small until you build up some tolerance.

Summary

Cultivating cannabis is complex and I wanted to provide you with a guide that could easily get the new grower started plus provide the experienced grower with some insight and a few tips to improve quality and yield.

Rather than just read a manual and connect the dots, I hope that I have encouraged you to make a connection with your plants and your environment. Don't get obsessed with details if you want to see the big picture. Plants are vibrational beings just like us. Learn to read them. Make that connection with nature. We are nature and nature rules.

We may not be able to implement all the rules and I know I have broken my own rules by writing this guide and putting myself out there on YouTube.

You can blow your brains out and spend a few million on a grow op or you can do it on a modest budget with a modest investment.

If you are interested in learning more about medical marijuana and its many benefits, pick up my guide The Medical Marijuana Guide. NATURE PHARMACY.

Learn how to make extracts and infusions like hash, hash oil, shatter tinctures, medicinal grade cannabutter and cannacoconut oil. The guide has many recipes for making marijuana edibles and much more. Plus you will learn about the nutritional and medicinal benefits of eating fresh raw marijuana.

Please take the time to write a review on kindle. It will help promote the book and get "The Medical Marijuana Growers Guide" noticed by others in need of a little guidance.

Peace and prosperity for all.

Chef Derek Butt.

Check out my Medical Marijuana Growers Guide playlist on YouTube. I will be producing more grow videos so don't forget to subscribe.

Bibliography

Chef Derek Butt. http://www.chefderekbutt.com

http://www.youtube.com/user/masterchefb1

Dr. Courteny. http://www.cannabisinternational.org/index.php

Dr. McAllister. http://www.cpmc.org/professionals/research/programs/science/sean.html

Dr. Hergenrather. http://medicalmarijuana.com/medical-marijuana-directory/listingDetails.cfm?lisID=739

Dr. Melamede. http://www.uccs.edu/~rmelamed/

Dr. Guzman. http://www.immugen.com/blog-for-cannabinoids-research/tags/Dr.-Manuel-Guzman/

Dr. Abrams. http://www.ucsfhealth.org/donald.abrams

Dr. Bearman. http://www.davidbearmanmd.com/publications.htm

Dr. Nagarkatti. http://pmi.med.sc.edu/PNagarkatti.asp

Dr. Mckuriya. http://mikuriyamedical.com/about/can_write.html

Pubmed http://www.ncbi.nlm.nih.gov/pubmed

Net Work

Check out my YouTube channel for instructional videos on everything covered in this book and then some. Also available to you are tips and tricks for cultivating medicinal marijuana. My videos are informative and entertaining.

I am also a composer, a musician, and a producer. I wrote the music in my videos. I studied music therapy at Capilano University and I think you will find my music, like most music to be therapeutic. Maybe not. I was kind of angry when I wrote it. It was therapeutic for me. Now that I am well medicated, I am looking forward to writing my next album.

http://derekbutt.com

YouTube Playlists:

Extracts and Infusions

Cannabis Infused Cuisine

Fresh Raw Marijuana

Medical Marijuana Growing Tips

From Seed To Harvest

Facebook, YouTube, Twitter, Linkedin Google+, chefderekbutt.com themedicalmarijuanaguide.com

The End

Made in the USA
San Bernardino, CA
27 February 2018